VIETNAM STUDIES

AIRMOBILITY

1961-1971

by

Lieutenant General John J. Tolson

DEPARTMENT OF THE ARMY

WASHINGTON, D.C., 1999

Library of Congress Catalog Card Number 72–600371

Reprinted by St. John's Press, Alexandria, Va.
978-1-944961-44-2
2016
www.stjohnspresspublishing.com

First Printed 1973—CMH Pub 90-4

For sale by the Superintendent of Documents, U.S. Government Printing Office Washington, DC 20402

Dedication

This study is humbly dedicated to the memory of the following airmobile battalion commanders who were killed in action during the period I commanded the 1st Cavalry Division in Vietnam:

Lieutenant Colonel Bob L. Gregory
Lieutenant Colonel Herlihy T. Long
Lieutenant Colonel Howard P. Petty
Lieutenant Colonel Robert L. Runkle

These gallant men—and all the honored dead of that war—will be always in the thoughts and prayers of their comrades-in-arms.

Foreword

The United States Army has met an unusually complex challenge in Southeast Asia. In conjunction with the other services, the Army has fought in support of a national policy of assisting an emerging nation to develop governmental processes of its own choosing, free of outside coercion. In addition to the usual problems of waging armed conflict, the assignment in Southeast Asia has required superimposing the immensely sophisticated tasks of a modern army upon an underdeveloped environment and adapting them to demands covering a wide spectrum. These involved helping to fulfill the basic needs of an agrarian population, dealing with the frustrations of antiguerrilla operations, and conducting conventional campaigns against well-trained and determined regular units.

As this assignment nears an end, the U.S. Army must prepare for other challenges that may lie ahead. While cognizant that history never repeats itself exactly and that no army ever profited from trying to meet a new challenge in terms of the old one, the Army nevertheless stands to benefit immensely from a study of its experience, its shortcomings no less than its achievements.

Aware that some years must elapse before the official histories will provide a detailed and objective analysis of the experience in Southeast Asia, we have sought a forum whereby some of the more salient aspects of that experience can be made available now. At the request of the Chief of Staff, a representative group of senior officers who served in important posts in Vietnam and who still carry a heavy burden of day-to-day responsibilities has prepared a series of monographs. These studies should be of great value in helping the Army develop future operational concepts while at the same time contributing to the historical record and providing the American public with an interim report on the performance of men and officers who have responded, as others have through our history, to exacting and trying demands.

All monographs in the series are based primarily on official records, with additional material from published and unpublished secondary works, from debriefing reports and interviews with key participants, and from the personal experience of the author. To facilitate security clearance, annotation and detailed bibliography have been omitted from the published version; a fully documented account with bibliography is filed with the Office of the Chief of Military History.

The author of this monograph, Lieutenant General John J. Tolson, has been involved with the airmobile concept since June 1939, when he participated in the first tactical air movement of ground forces by the U.S. Army. Participating in all the combat jumps of the 503d Parachute Infantry Regiment during World War II, he became an Army aviator in 1957, and later served as Director of Army Aviation and Commandant of the U.S. Army Aviation School. From April 1967 to July 1968 he served as Commanding General, 1st Cavalry Division (Airmobile), Vietnam. General Tolson is now Deputy Commanding General, Continental Army Command.

Washington, D.C.
15 November 1972

VERNE L. BOWERS
Major General, USA
The Adjutant General

Preface

The purpose of this study is to trace the evolution of airmobility in the U.S. Army. The integration of aircraft into the organic structure of the ground forces is as radical a change as the move from the horse to the truck, *and the process is only beginning.*

Because this change is not the product of one man or one small group of men but rather a fortunate confluence of technology, tactics, and imagination, proper credit to every responsible individual is impossible. I have tried to identify some of those people who made a major contribution throughout the years. I apologize to those people whose names have been omitted either because of oversight or lack of space.

Although Vietnam was the first large combat test of airmobility, air assault operations in Southeast Asia would not have been possible without certain key decisions a decade earlier. This study attempts to trace the most important milestones which led to the eventual formation of airmobile divisions.

It would be impossible in a single volume to adequately describe every airmobile operation in Vietnam during the years 1961–1971. Therefore, only selected operations have been chosen as examples of different airmobile tactics. Many of these were selected because of the author's personal knowledge. Another author might have selected different operations.

I believe I'd be remiss in this account if I were not candid with the reader on some of the pros and cons of airmobility. Thus, throughout the text, I have inserted comments that are intended to broaden the reader's view of this issue.

This study is aimed at a broad audience, some of whom may only have a passing familiarity with Army aircraft systems. These systems are pictured in an appendix with appropriate data on each.

Over 300 major source documents were reviewed during the preparation of this study. For the serious student we have identified a rich vein that, as yet, has not been deeply mined. Thousands of important stories, yet untold, lay buried in these pages. Length has permitted us to extract only a few.

I wish to thank the many senior officers who went out of their way to contribute their own special comments for this monograph and the officers who wrote special studies as basic reference material for the monograph. I have drawn extensively on two such studies prepared by Lieutenant Colonel John R. Galvin of Combat Developments Command and Major Bobby D. Harber of the U.S. Army Aviations Systems Command. Because the published version of my monograph contains no documentation, the extent of my indebtedness to these two officers is not readily apparent. Major Harber's manuscript was subsequently published by the U.S. Army Aviation Systems Command under the title, *Logistical Support of Airmobile Operations, Republic of Vietnam, 1961–1971*.

Finally, I must recognize the two officers who were primarily responsible for researching, drafting, and compiling this volume. Colonel James J. Brockmyer, who was my senior assistant, has been associated with airmobility for more than two decades and was the editor of the test report of the 11th Air Assault Division. CW–2 Charlie M. Montgomery, who researched and typed the multiple drafts, was a special assistant to General Westmoreland for four years in Vietnam. These two officers must share with me any credit (or blame) that this study might generate.

Airmobility is no panacea; it brings with it many unique problems as well as unique capabilities. It is hoped that this study will give the reader some insight into both of these areas.

Washington, D.C.
15 November 1972

JOHN J. TOLSON
Lieutenant General, U.S. Army

Contents

Charts

Maps

Illustrations

All illustrations are from Department of Defense files.

AIRMOBILITY, 1961-1971

CHAPTER I

The Growth of the Airmobile Concept

The First Airmobile Units in Vietnam

On 11 December 1961 the United States aircraft carrier USNS *Card* docked in downtown Saigon with 32 U. S. Army H–21 helicopters and 400 men. The 57th Transportation Company (Light Helicopter) from Fort Lewis, Wash., and the 8th Transportation Company (Light Helicopter) from Fort Bragg, N. C., had arrived in Southeast Asia. This event had a two-fold significance: it was the first major symbol of United States combat power in Vietnam; and, it was the beginning of a new era of airmobility in the United States Army.

Just twelve days later these helicopters were committed into the first airmobile combat action in Vietnam, Operation CHOPPER. Approximately 1,000 Vietnamese paratroopers were airlifted into a suspected Viet Cong headquarters complex about ten miles west of the Vietnamese capitol. The paratroopers captured an elusive underground radio transmitter after meeting only slight resistance from a surprised enemy. Major George D. Hardesty, Jr. of the 8th Transportation Company and Major Robert J. Dillard of the 57th could report that their units had performed outstandingly under their first baptism of fire.

The events of December 1961 prefaced a decade of unparalleled growth of airmobility. But they also were a culmination of many decisions during the preceding decade which allowed the President of the United States to exercise this option in support of the Government of Vietnam.

The Growth of the Concept

In its broadest sense, the airmobility concept envisages the use of aerial vehicles organic to the Army to assure the balance of mobility, firepower, intelligence, support—and command and control. The story of airmobility in the Army is inextricably interwoven with the story of Army aviation. Airmobility could be viewed as the warp of new tactics and techniques with organic aviation providing the woof of skills and technology to make the whole cloth.

The airmobility concept was not a product of Vietnam expediency. It would probably not be practical to make a record of every decision which formed a part of the airmobile concept. It certainly had its roots in both the airborne techniques of World War II and the early doctrine for organic aviation for ground forces for that era. However, from a practical viewpoint, *the* embryonic decision can be said to be the Army's move to form twelve helicopter battalions on 21 August 1952. This key decision was made to commit the Army to airmobility even though a practical troop-carrying helicopter was still unproven and the helicopter tactics and techniques existed only in the minds of a handful of men. Though the twelve battalion goal was overly optimistic at that time, it dramatized how the impact of Korean experience had influenced Army planners.

In Korea the Army had learned that the difficult terrain in that land mass and the numerical superiority of the enemy combined to provide the communist with an advantage that was not easily balanced by the qualitative superiority of the American soldier and his supporting firepower. Many thoughtful officers had watched the little observation helicopter—turned ambulance—flit up and down the steep hills with effortless agility. It was not hard to envision the possibilities inherent in hundreds of larger machines carrying combat troops up and over those deadly slopes. The Marines had already demonstrated the possibility of a small unit being carried into combat and the helicopter itself was beginning to mature, with more power and more dependability.

After Korea many senior commanders restudied the lessons of that war and compared actual campaign operations with hypothetical airmobile operations under the same conditions. Various Army aviators and members of the helicopter industry were keeping a close watch on the French and British helicopter operations in Algeria and Malaysia.

The mid-fifties were gestation years for new tactics and technology. Major General James M. Gavin's article in *Harper's* magazine (April 1954) "Cavalry, and I Don't Mean Horses!" was indicative of the sort of vision some enthusiasts had for the future. The article was an unofficial summation of several staff studies prepared in his office while he was G-3, Department of the Army. During this period I was General Gavin's Director of Doctrine and Combat Development, and through his Deputy, Major General

Paul D. Adams, he had tasked my shop to design new hypothetical cavalry organizations around the potential of the helicopter. These new units were to perform the traditional missions of horse cavalry using a third dimension and a ten-fold increase in speed. Not entirely by accident, one of the key action officers was a young Major Edwin C. ("Spec") Powell, later to become Director of Army Aviation.

The growing helicopter industry was groping for direction from the military. At the same time, its engineers were pushing design proposals to potential military customers who weren't at all sure of their requirements. Many new approaches to vertical (or near vertical) flight were being tested as a replacement for, or an extension of, the helicopter. For the first time, the Army placed major Research and Development funds in aëronautic programs.

There were precious few guidelines for airmobile doctrine in the early fifties and none had the formal stamp as official U. S. Army policy. Recognizing this deficiency, General Gavin decided to plant the seed in the Army School System. Through the good offices of Major General Robert M. Young, G–1 of the Army, and Major General Joseph H. Harper, Commandant of the Infantry School, I was ordered to Fort Benning with instructions to develop tactical doctrine for the combat employment of helicopters.

As Director of the Airborne Department, I was able to broaden its charter and change the name to the Airborne-Army Aviation Department. A new Airmobility Division was established as the focal point for doctrinal innovation. We started gathering the necessary people and equipment to develop procedures, organizational concepts, and matériel requirements for airmobility. A helicopter company, equipped with H–19's and, later, H–34's was placed at our disposal for field experiments. The company commander, Major William C. Howell, Jr., was to become President Eisenhower's personal helicopter pilot and, after retirement, curator of the U. S. Army Aviation Museum at Fort Rucker.

The results of a few months of concentrated activity were climaxed with the publication of a new field manual, FM 57–35, "Army Transport Aviation–Combat Operations." The basic tactics and techniques described in this manual stood the test of time and would be vindicated in the tests of the 11th Air Assault Division and in Vietnam.

The first formal military requirement for an aerial weapon for Army helicopters was initiated during this same period at Fort Benning and, subsequently, approved through the chains of command by Department of the Army. The requirement specified a light-weight, simple, flexible turret gun to be provided for all Army transport helicopters. Tactical experiments had proven the obvious requirement for some type of suppressive gunfire to be delivered by transport helicopters during the critical approach phase of a combat assault. Unfortunately, the light turret gun was never developed as conceived, but the Ordnance Corps latched onto the requirement as the justification of expenditures in the initial development of helicopter armament.

The Armed Helicopter

During this same time frame, a small group of pioneers, with the blessing and backing of Brigadier General Carl I. Hutton, Commandant of the Aviation School at Fort Rucker, Alabama, were making the first crude experiments with armed helicopters. Scrounging discarded hardware from the other services' junkyards, they improvised various combinations of machine guns and rockets in the face of ridicule and outright opposition. At the same time they began to develop tactics and techniques which they felt would prove useful when the technology of helicopter armament caught up to the theory.

The guiding genius for much of this development was a colorful officer, Colonel Jay D. Vanderpool. Colonel Vanderpool was Chief of the U. S. Army Aviation School's Combat Development Office, which was under the direct supervision of the Assistant Commandant. This arrangement was fortuitous since we could make all the assets of the School available to his project if required. In June 1956, with two officers, two enlisted men, and unbounded enthusiasm, Colonel Vanderpool went to work without a charter, without money and, by explicit direction, without publicity.

With borrowed personnel from the Department of Tactics, Colonel Vanderpool formed a "Sky-Cav" platoon which became notorious for its hair-raising demonstrations of aerial reconnaissance by fire. By mid-1957 this provisional unit, redesignated Aerial Combat Reconnaissance Platoon, had somehow acquired two H-21's, one H-25, and one H-19 armed with a wondrous variety of unlikely weapons. Colonel Vanderpool and his "hoods" were to see their efforts officially recognized when the Aerial Combat Reconnaissance Platoon became the nucleus of the 7292d Aerial Combat Reconnaissance Company (Provisional) with an approved Table of Distribution sanctioned by the Department of the Army on 25 March 1958. In subsequent work, Colonel Vanderpool and his group developed, on paper, organizations for an "Armair" troop, squadron, and brigade. As the Assistant Commandant of the Army Aviation School, I was present when Colonel Vanderpool gave a briefing to Major General Hamilton H. Howze, the first Director of Army Aviation, on the organization of an "Armair *Division*."

Staff Plans an Army Aircraft "Family"

Near the end of the 1950's, the Army staff recognized the overwhelming need to get its "aviation house" in order. It was becoming big business. Too many programs were going in too many directions. Many Army agencies had separate responsibilities and were pursuing their own program in happy isolation of the rest. The Transportation Corps and Signal Corps had reluctantly brought their test activities to Fort Rucker with the Army Aviation Test Board, but there were still many fragmented efforts at many echelons. The Army seemed uncertain of its goals; Congress was unhappy with some of the budget justifications; and, the aviation industry was uncertain.

At the end of June 1959, the Army had 5,500 aircraft, most of them substandard, with a requirement for 6,400. Taking into account normal losses, wearout, and attrition, the aviation picture was very poor for 1970. The greatest need was for a light observation aircraft to perform the "bread and butter" mission of Army aviation.

One of the most important milestones during this period was the decision to develop the XH–40 Bell Utility Helicopter and to power it with a turbine engine. Although designed as an aerial ambulance, it was recognized even then that this machine might turn out to be the most useful aerial platform ever put in production.

In October 1959 the Army Chief of Research and Development, Lieutenant General Arthur G. Trudeau, initiated an Army Aircraft Development Plan to advance firm guidance for research and development efforts during the next decade. He felt the plan would bridge the gap between the Army and U. S. Air Force capabilities and give the Army the initiative in the manned aircraft field. Industry was expected to contribute a major effort.

To implement the Army Aircraft Development Plan, the Office of the Chief of Research and Development prepared three broad development objectives which they called Army Study Requirements. These study requirements forecast a need for light observation, manned surveillance, and tactical transport aircraft. They were presented to all members of industry on 1 December 1959 at

Fort Monroe, Virginia. The requirements spelled out enough information, along with available parallel U. S. Air Force and U. S. Navy studies, to enable industry to explore technical approaches to meet the requirements.

The Rogers Board

On 15 January 1960 the Army Chief of Staff established the Army Aircraft Requirements Review Board chaired by Lieutenant General Gordon B. Rogers, the Deputy Commanding General of the Continental Army Command, to consider the Army Aircraft Development Plan and to review the industry proposals. Other members of the Board were Major Generals Robert J. Wood, Hamilton H. Howze, Thomas F. Van Natta, Alva R. Fitch, Richard D. Meyer, and Ernest F. Easterbrook; and Brigadier Generals Lawrence J. Lincoln, William M. Thames, Jr., and Clifton F. von Kann. The Board was supported by the Army general staff and elements of the Continental Army Command. Colonel Robert R. Williams, then Chief of the Airmobility Division, Office of the Chief of Research and Development, was the guiding genius behind the formation of the Rogers Board and Secretary (without vote). As the Deputy Director of Army Aviation, I also served on the Board without vote.

On 1 February 1960, forty-five companies submitted 119 designed concepts as their solutions to the problems posed by the Army Study Requirements. These concepts were evaluated in two phases—a technical evaluation under the direction of the Chief of Transportation from 1–15 February and an operational evaluation from 16–28 February under the direction of the Office of the Chief of Research and Development.

It is important to note that this was the first time that most of the major aircraft companies took official notice of the aviation potential within the Army. Before the Rogers Board, only the helicopter manufacturers and a few light airplane companies had any comprehension of the Army's requirements.

After the evaluation period, the Rogers Board met at Fort Monroe from 29 February to 6 March. During that time it reviewed the Army Aircraft Development Plan, discussed roles and missions of Army aviation, projected Army funding, assessed combat surveillance requirements, and detailed procurement plans.

The Rogers Board made recommendations regarding three types of aircraft—observation, surveillance and transport. In the observation field it advocated conducting an immediate design competition to develop a new helicopter, selecting at least two designs for full development, and competitive tests. The selected design would begin production in Fiscal Year 1964 and eventually phase out the existing observation aircraft, the L–19, H–13, and H–23. In the surveillance field, the Rogers Board believed that more testing on sensory devices, data link, and intelligence processing were necessary before military characteristics could be prepared for a penetration surveillance aircraft. Pending these studies, the Board recommended a new aircraft be developed with an operational target date by 1970. In the area of transport aircraft, the Board determined that more specifics were needed on exact requirements for Army airlift to support contingency plans and a program was needed to provide a vertical or short-take-off-and-landing long-range replacement for the Chinook and Caribou in the early 1970's.

Two other recommendations were made by the Board which are not generally well-known. The Board recommended the establishment of a policy to replace each model of aircraft at least every ten years or sooner if warranted by operational requirements or technological advances. It also recommended that Department of the Army and Continental Army Command prepare an in depth study to determine *whether the concept of air fighting units was practical* and if an experimental unit should be activated to test its feasibility.

On 19 March 1960 the Army Chief of Staff approved the Rogers Board recommendations with implementation for planning purposes and assigned various staff agencies primary responsibility to carry out these recommendations. Regarding air fighting units, he charged Deputy Chief of Staff for Operations with the responsibility for preparing recommendations in this area.

The importance of the Rogers Board has been somewhat obscured by the later Howze Board and tests of the 11th Air Assault Division. However, it was a remarkable milestone in Army airmobility. It set forward a chain of actions which had a profound effect on later concepts.

With historical hindsight, it is apparent that the scope of the 1960 Rogers Board review was limited. It obviously did not constitute a major advance in tactical mobility for the Army. But in comparison with the advances made during the 1950's, the Board's objectives, if obtained, would have represented a substantial gain in mobility through the use of aviation. The Rogers Board provided essential aviation guidance for development, procurement, and personnel planning.

The work of the Rogers Board was symptomatic of a renaissance throughout many segments of the Army—in its schools and its fighting units. As an example of the latter, Lieutenant Colonel Russell P. Bonasso, the Aviation Officer of the 101st Airborne Division, made what could be called a drastic organizational proposal in 1960. After studying the fragmented aviation assets in the division, he briefed the Commanding General, Major General William C. Westmoreland, on the advantages of centralized control. At the conclusion of a lengthy discussion, General Westmoreland authorized the formation of the 101st Combat Aviation Battalion (Provisional)—the first such organization in the Army.

Army-Air Force Differences

The early development of the airmobile concept was not without controversy. As the direct result of the experiences in Korea and the disagreements over Army aviation between the Air Force and the Army, Secretaries Frank Pace, Jr. of the Army and Thomas K. Finletter of the Air Force signed a special Memorandum of Understanding dated 2 October 1951. Under the terms of this document organic Army aircraft would be used by the Army "as an integral part of its components for the purpose of expediency and improving ground combat and logistics procedures within the combat zone." Detailed functions under the exclusive control of the ground force commander which might be performed by Army organic aircraft were spelled out in this Magna Carta of Army aviation. Predictably this did not settle the so-called roles and missions issues between the Services. Other memoranda were to follow.

The Air Force witnessed the rapid growth of Army aviation during the 1950's. From 668 light airplanes and 57 light helicopters that comprised the Army inventory on 30 June 1950, it saw the Army acquire over 5,000 aircraft of fifteen different varieties by 1960. The Air Force had watched the Army become the acknowledged leader in vertical flight and ground effects machines. Against strenuous objections, the Army had "borrowed" three Air Force T-37 jets for testing and a number of Army aviators were being qualified in various transonic aircraft from other Services and the North Atlantic Treaty Organization. Worst of all, the Air Force found itself with a growing number of "technologically unemployed" pilots and they heard themselves described in front of a Congressional committee as the "silent silo-sitters-of-the-seventies."

A briefing of particular significance took place in the late afternoon of 12 December 1960. General Thomas D. White, Chief of Staff of the Air Force, had requested this briefing on the future Army aviation program and, as the Deputy Director of Army Aviation, it was my duty to present the formal portion of the briefing in the presence of General George H. Decker, the Army Chief of Staff, and many senior officers of both Services.

At that time the focus in the Army was on the nuclear battlefield. Organic aviation was viewed by the Army as the best means of maintaining combat operations in an area characterized by great depth and frontage with the dispersion of many small self-contained units. The major threat was viewed as a sophisticated enemy attacking with masses of armor on the plains of Europe. Counter-guerrilla warfare at that time was viewed as a secondary mission. Nevertheless, the early planners in airmobility perceived that one of the automatic fallouts in organizing the Army for greater airmobility would be much greater capabilities in the lower spectrums of warfare.

The Army had already made a decision to replace its light fixed-wing observation airplane and the two light observation helicopters with a single light turbine helicopter. This had been a fundamental decision of the Rogers Board the year before and was another example of the Army's commitment to the turbine engine. The early Bell XH–40 had been standardized as the HU–1 and was envisioned then as the replacement for the L–20 utility airplane and the H–19 utility helicopter. Further growth versions of the Bell machine were planned to replace the bulk of the missions then performed by the Sikorsky H–34 and the Vertol H–21. The Vertol HC–1B Chinook was then on the drawing boards to replace the piston powered Sikorsky H–37.

To provide background for understanding the question and answer period which followed the briefing I must digress to comment on the Mohawk and the Caribou aircraft.

Since its inception as a joint Army and Marine program, the OV–1 Mohawk had been a center of controversy. Actually the Army and Marine requirements were never compatible and compromises were made that suited neither. From the Army's viewpoint, the original design was compromised by shipboard requirements and other specific Marine specifications which had little application for an Army observation aircraft. From the Marine viewpoint, they were looking for a fixed-wing replacement for the old Cessna light observation aircraft and they did not require sophisticated sensor systems which they planned to carry on other aircraft. As it turned out, the Marines dropped out of the development program before the first prototypes were ready for flight.

Because the Mohawk was an exception to the Secretary of Defense's memorandum on weight limitations for Army aircraft and because it had inherent capabilities for armament, the Air Force had opposed its development from the beginning. There is no doubt that certain Army extremists viewed the Mohawk as the "nose of the camel within the tent of tactical air support." The Army was to suffer for their enthusiasm for years to come. The manufacturer, Grumman, did not help by publishing carefully placed brochures which showed the Mohawk in a variety of attack roles.

The Mohawk was originally designed as a visual reconnaissance aircraft with better survivability than the L–19 of Korean vintage. In addition, it was to have an integral camera system for spot photo coverage. Above all, it was to land and take off in the same distance as the L–19 which it was to supplement. It was not long, however, that "improved" versions of the Mohawk were visualized carrying sophisticated sensor systems developed by the Army surveillance agencies such as infra-red and side-looking radar. Weight, space, and power provisions had not been made for these systems in the original design. As a result, the gross weight increased and performance declined. These growth versions of the Mohawk were coming off the drawing board before the first "A" model had even been tested throughout its entire flight envelope and subsequent tests were to prove that major engineering modifications to both wing and power plants would be necessary in the latter versions. Furthermore, the addition of all this sophisticated sensor equipment not only raised the unit cost significantly but, in the view of many, watered down the Army contention that this was truly a front line low echelon aircraft.

Turning to the Caribou, this aircraft came into the Army by the sheer persistence of a few perceptive representatives of deHaviland Aircraft Corporation of Canada and a few equally perceptive Army aviation enthusiasts who recognized the natural affinity of the deHaviland aircraft to the Army environment. Starting with the Beaver—which was designed as a "bush pilot's dream," and progressing to the reliable, but somewhat clumsy Otter—the Caribou seemed a natural to those Army planners who saw a significant gap between the Army's organic capability and the C–130 Air Force transport.

The Caribou was the second exception by the Secretary of Defense to his memorandum on the weight limitations for Army fixed-wing aircraft. Its very size threw down a gantlet to the Air

Force strategists. It could carry as much as the venerable C-47 and land and take off of remarkably short unimproved airstrips. It had a rear end loading door which facilitated rapid cargo handling and was relatively easy to maintain and operate. The Air Force saw this 32 passenger aircraft as a second invasion of its domain and the projected rise of the Army's inventory at a time that the Air Force aircraft inventory was declining was bound to cause more heat than light.[1]

The Caribou proved to be an extremely useful addition to the United States aircraft inventory and did yeoman service in Vietnam. Later it would be transferred to the Air Force in 1966 as part of another Army-Air Force agreement, but in the early 1960's the Caribou and the Mohawk were the two major symbols of Army-Air Force disagreement and more time was devoted to these systems than to the entire airmobility concept itself.

The question and answer period that followed the briefing began with a discussion about the Russians' effort in the helicopter field and the Russians' method of organizing their aviation. General White then turned to the growth of aviation in the United States Army and got to the heart of the meeting. To the obvious dismay of some Air Force staff officers, he said,

> I don't disagree with the Army concept. Obviously there is always going to be the clash of budgetary impingements. Your idea of supplying yourselves and this "Sky Cav," I can't quarrel with. But what worries me and must worry you for other reasons is what does the future hold? Two of these airplanes are probably as big as our bombers were eighteen or twenty years ago. What does the future hold? How do we get along, because the air does clash at a certain point? How do we work this out?

General Decker indicated that Army aviation would take up where the Air Force left off and that he was perfectly happy to leave it that way—provided the Air Force provisioned the necessary airlift and tactical air support to complement the Army's need. The conversation then shifted to the type of close air support that the Army felt was best. General Decker espoused a single-purpose,

---

[1] The very existence of the Caribou in the Army inventory changed one Air Force program—the Fairchild C-123 "Provider." This latter aircraft was scheduled to be phased out of the active Air Force inventory in 1961, with a few programmed to the reserves and the remainder to be declared surplus. When Secretary of Defense Robert S. McNamara suggested that the Air Force turn over the C-123's to the Army (to train on this type aircraft prior to the receipt of the Caribou), the Air Force suddenly discovered new and pressing Air Force requirements for the C-123. Thus the Army Caribou protagonists not only pushed the Caribou into being, but—incidentally—saved the C-123 for much-needed duty in Vietnam.

simple combat airplane for close support, while General White defended the Air Force viewpoint that a multi-purpose aircraft could do the job better and also perform as an air defense weapon. The rest of the conversation centered about the Army's testing of certain jet aircraft for deep surveillance. Finally, the problems of command and control of two distinct systems—Air Force and Army—were discussed.

General White and General Decker closed the session with a resolution to have Major General David A. Burchinal and Major General Barksdale Hamlett work on more specific agreements between the Air Force and the Army. The two Chiefs of Staff closed the session with the following remarks:

> General White: We waste more money; we waste time; we waste energy; and we waste dispositions. I doubt that we will solve all our problems but I am certainly for solving as many of them as we can.
>
> General Decker: Well, I am, too. We will go halfway or even more. I will say this, that there is a group around this city not confined to the Army, who think the Army is not doing itself justice for coming out and making a pitch to take over some of these things such as tactical air support. I won't agree with them. I think as long as the Air Force has the mission and can do it they ought to do it. We don't want to take it over, but there is that kind of pressure as you probably know. I want you to know that that is not my position.

The Air Force and the Army differences brought out in the briefing just described were not to be entirely solved during the next decade, but many problems between the services disappeared in the crucible of combat. Working agreements between those actively engaged in battle were developed and standardized. But the fundamental differences remained between those who viewed aviation as an end in itself and those who saw aviation merely as a means to free the ground soldier from the "tyranny of terrain."

Reviewing the Army's position in 1960, a full year before the first helicopter companies landed in Saigon, it should be noted that the Army was well along its way in airmobility. At the time it had the following basic objectives: each division to have the capability of moving at least a company of Infantry by its organic airlift; aerial surveillance to match the firepower of the unit at each echelon; rapid purification of the aircraft inventory to reduce the types of aircraft to the minimum; acquisition of a limited number of flying cranes; and, increased logistical capabilities as represented by the Caribou and Chinook team and any possible successor. It is important to view these objectives in their proper time frame for they were not necessarily geared to the prospect of an increased involvement in Vietnam, but rather to a new Army structure, worldwide. The planners at that time felt certain that if they provided increased airmobility for most Army units it would be useful for the major contingencies that could be forecast.

Vietnam Fleet Expands

Early in 1961 General Maxwell D. Taylor, who was then the Military Advisor to the President, made a situation survey visit to Southeast Asia. During his visit he felt that the lack of adequate roadnets, lines of communication, and means of mobility contributed heavily to the government's problems in South Vietnam. As a result of General Taylor's recommendations, President John F. Kennedy approved a more active support program to South Vietnam to assist in the fight against the communist-directed Viet Cong. Generally, the support included the establishment of a joint headquarters for directing the program; increasing the number of U. S. advisors for the South Vietnamese armed forces; and additional support through Army Aviation, communications units, and Navy and Air Force units. This began a chain of events culminating in the arrival of the helicopter units mentioned at the beginning of this study.

The first few helicopters to arrive "in country" were based at Tan Son Nhut and provided support to all the Army of the Republic of Vietnam units they could reach. Naturally, this support was based on operational priorities–the average Army of the Republic of Vietnam infantry unit saw very few helicopters in its day-to-day operations. This was, however, a period of innovation and of "trial-and-error" schooling for the U. S. pilots and crews, and also for the Viet Cong.

There is one hair-raising tale of the H–21 pilot who put down in a landing zone only to see a squad of Viet Cong step from the woods and open fire on him at point blank range. Fortunately for the pilot and crew, the Viet Cong squad leader had firmly embedded in the minds of his men the necessity to "lead" the helicopter with their fire. The amazed pilot watched the squad pour their fire into the ground 20 yards in front of him. He took off without a single hit. This kind of strange tale in a sense characterizes the early days–you might say the bailing wire days–of airmobility in the Vietnam War.

Following the 57th and 8th, the 93d Transportation Company (Light Helicopter) arrived off the coast of Vietnam in January 1962. Ten miles out in the South China Sea from DaNang, the aircraft were flown off the carrier deck of the USNS *Card* to Da

Nang Air Base. This unique delivery was accomplished without serious incident even though ceilings were down to 100 feet over the ocean. These three aviation companies experienced seemingly insurmountable difficulties because of a critical shortage of engines and deterioration of rotor blades and aviation equipment due to high humidity. Nevertheless, they continually overflew their programmed flying hours and exceeded aircraft availability normal rates.

To some extent the support gap was bridged upon arrival from Fort Riley, Kansas, in January 1962 of the 18th Aviation Company (U-1A Otter). These aircraft were spread throughout the four corps areas to provide a utility supply net throughout the length of the country. Most of their missions involved delivery of aircraft parts and supplies to rotary wing aviation units that were widely separated from their support elements.

The first Marine helicopter squadron arrived in country in April 1962 and was established at the old French base at Soc Trang in the Mekong Delta. In June and July of that year the Marines swapped bases with the 93d Transportation Company at Da Nang because of the greater capability of the Marine H-34 helicopters to operate in the higher elevations of the northern region.

To provide better command and control of the Army's growing fleet, the 45th Transportation Battalion was deployed to Vietnam in early 1962 from Fort Sill, Oklahoma, and assumed command of the three Army helicopter companies and the fixed-wing Otter company. Shortly thereafter two more light helicopter companies, the 33d and the 81st, were deployed and also came under the command of the 45th Transportation Battalion.

The first of "a long line of Hueys" arrived in Vietnam as part of the 57th Medical Detachment (Helicopter Ambulance) in early 1962. They were shortly followed by the 23d Special Warfare Aviation Detachment equipped with OV-1 Mohawks to provide reconnaissance and photographic coverage in support of ARVN forces.

Much of this early effort was classified. As a result, many Army aviators felt a bit frustrated that they were being kept "under wraps" at the same time they read of Marine exploits around Da Nang in the daily newspapers. Nevertheless, their operations *were* receiving official attention at the highest levels.

The Events Leading to the Howze Board

While the first Army aviation units were deploying to Vietnam, settling in-country and making their first tentative tests in combat, events in Washington were occurring which would have a profound influence on the future of airmobility.[2] Some of the events which greatly influenced an increasing use of airmobility included the reorganization of the Army divisions, an increase in the number of divisions, the Berlin buildup and call up of reserves, insurgency operations in general, and additional requirements for Army-type aircraft in military assistance programs throughout the world. The Army found that the requirements for its aircraft outpaced current procurement and deployment of aviation units depleted the inventory in the Continental United States.

In late September 1961 Secretary of Defense McNamara reviewed the Army aviation plans and indicated tentative decisions concerning procurement of Army aircraft and the number of Army aviation companies. Uneasy over his tentative decision in this matter, he sought further information. On 4 October he conferred with General Clyde D. Eddleman, Army Vice Chief of Staff, Brigadier General von Kann, Director of Army Aviation, and Major James J. Brockmyer, Army Aviation Action Officer. During this conference he sympathetically questioned the Army program, especially the Bell utility helicopter program, stating that he felt the procurement was strung out too many years. If the Army really needed these helicopters and its design offered many advantages over the current models, he felt the early procurement of large numbers would best serve the national interest.

As a result of this conference and the previous study effort of his own staff, Mr. McNamara sent a short memorandum to the Secretary of the Army on 5 October requesting a study of all Army aviation requirements. As part of this study, he desired an Army evaluation of his previously proposed cuts in the Army aviation program. He asked for a progress report by 20 October and a final report by mid-November.

General von Kann, as Director of Army Aviation, began to compile his own report of the entire Army aviation program and a forecast for requirements for aircraft through 1970. By the time Mr. McNamara's memorandum reached the Army Staff, this report was well underway. General von Kann's report and the reclama of the proposed reduction in Army aircraft procurement constituted the Secretary of the Army's answer to the Secretary of Defense.

---

[2] After living with the early problems of airmobility for six full years, it was a disappointment to me that I would be out of the mainstream of airmobility actions for the next exciting two years. I was promoted out of my job as Deputy Director of Army Aviation in the spring of 1961 and served as Chief, MAAG, Ethiopia until 1963. I was then fortunate to return to the Pentagon as Director of Army Aviation.

Analysts in the Office of the Secretary of Defense reviewed the Army's submission during January and February 1962. Their review was extremely critical of the Army's so-called caution. Key action officers prepared a draft for Mr. McNamara which would have a long-range effect on the structure of the Army.

On 19 April 1962 Mr. McNamara sent a now famous memorandum to the Secretary of the Army in which he stated he felt the Army's program was dangerously conservative. He believed the procurement program fell short of meeting requirements and failed to give a rationale for an optimum mix of aircraft types. The staff paper supporting the memorandum listed three major criticisms of the Army aviation program. First, OSD did not think the Army had fully explored the opportunities offered by technology to break their traditional ties to surface mobility. Second, they claimed that air vehicles operating close to the ground offered a possible quantum increase in effectiveness. Third, they stated that air transportation, all things considered, was less costly than rail or ship transportation even in peacetime, and wartime urgency would make it even more important. Therefore, the OSD staff stated that the Army needed a major effort to exploit the aeronautical potential and increase its effectiveness vis-a-vis ground transportation systems.

The Secretary believed the Army needed to re-examine its aviation requirements with a bold "new look" at land warfare mobility. Though previous studies had produced doctrinal concepts, action was needed to carry the concepts into effect. Studies of air vehicles were needed also. He directed the re-examination of tactical mobility requirements divorced from traditional viewpoints and past policies, and free from veto or dilution by conservative staff review. He felt the Army should be willing to substitute airmobile systems for traditional ground systems if it would improve capabilities and effectiveness. The final objective of the re-examination was to recommend action which would give the Army the maximum attainable mobility in the combat area within the bounds of aeronautical technology, within alternative funding levels.

The Secretary of Defense directed that the results of re-examination be presented in terms of cost effectiveness and transport effectiveness, and that full use be made of field tests and exercises. Mr. McNamara urged Secretary of the Army Elvis J. Stahr, Jr. to give the matter his personal attention and, in a most unusual departure from accepted procedure, suggested the following individuals to manage the Army's effort: Lieutenant General Hamilton H. Howze, Brigadier Generals Delk M. Oden, Walter

B. Richardson, and Robert R. Williams; Colonels John Norton and Alexander J. Rankin; and Mr. Frank A. Parker Jr., President of the Research Analysis Corporation, Dr. Edwin W. Paxson of the RAND Corporation, and Mr. Edward H. Heinemann, a well-known aviation consultant.

Secretary McNamara summed up the emphasis he placed on the Army reexamination of tactical mobility this way: "I shall be disappointed if the Army's reexamination merely produces logistically oriented recommendations to procure more of the same, rather than a plan for employment of fresh and perhaps unorthodox concepts which will give us a significant increase in mobility."

This benchmark in airmobility history resulted from the fortunate confluence of several trends: first, the personal dissatisfaction of the Secretary of Defense with the Army's failure to exploit the potential capabilities of airmobility; secondly, an undeniable attitude of many office of the Secretary of Defense civilian analysts who looked upon the service staffs and most officers as reluctantly being dragged into the twentieth century; third, there was a nucleus of Army aviation oriented officers both in the office of the Secretary of Defense staff and Army Staff who recognized the possibility of capitalizing on Mr. McNamara's attitude to sweep aside ultraconservative resistance within the Army itself. Finally, there was an opportunity to present to the Secretary of Defense for his signature directives that would cause the Army to appoint an evaluation by individuals known for their farsightedness and to submit recommendations directly to the Secretary of Defense in order to avoid intermediate filtering. For the record, it should be noted that General Howze knew nothing of this background maneuvering and would have sternly protested had be been aware.

The Army Vice Chief of Staff immediately alerted the General Staff of Secretary McNamara's requirements. The staff suggested the appointment of a board of general officers and selected civilians with subordinate working groups similar to the organization of the Rogers Board. Department of the Army directed General Herbert B. Powell, Commanding General, Continental Army Command, to establish the Board which would have three suspense dates to meet. On 10 May 1962 the Board would submit an outline plan of how it would conduct its review, including an estimate of funds required. Beginning 1 June the Board would submit monthly progress reports and by 24 August a final report incorporating a recommended program for development and procurement of Army aircraft. Department of the Army directed other agencies to provide appropriate support and assistance as required to Continental Army Command and the Board.

The Howze Board

Within a week after Secretary McNamara's memorandum of 19 April, Continental Army Command appointed General Howze, Commanding General of the Strategic Army Corps and of the XVIII Airborne Corps and Fort Bragg, as president of the *ad hoc* U. S. Army Tactical Mobility Requirements Board to conduct a reexamination of the role of Army aviation and aircraft requirements. On 25 April Continental Army Command contacted General Howze to alert him of the job to be done, outlined preliminary instructions, and provided pertinent details. The formal directive dated 3 May was not issued until the Board had begun work. In fact, General Howze did not read it until 5 May. Continental Army Command warned General Howze that the deadlines were extremely demanding. On 25 April, the same day on which General Howze was notified that he was to form the Board, General Decker, Chief of Staff of the Army, informed his staff of the establishment of the Board and outlined the Board's purpose and function, composition, direction and control, and administrative procedures. His directive formalized the verbal instructions previously given.

General Howze initial reaction was that the magnitude of the job and the consideration of avoiding false starts might prohibit the Board from accomplishing all the assigned tasks by the deadline. For example, he believed that exercises and troop tests should be continued beyond that time frame to further refine requirements for aircraft and of the matériel. However, he felt that within the time limitation the Board could arrive at most of the essential conclusions and provide a conservative estimate of the proper program on which the Army could embark without wasting assets.

General Howze was free to convene portions of the Board at installations other than Fort Bragg if convenient. He could organize the Board as he saw fit. Within the first few days of May, aviation units and support elements awaited his bidding, funds had been earmarked, and instructions had been issued to insure that supply and maintenance priorities were clearly understood throughout the Army. General Howze was free to deal directly with the Department of the Army, the Department of Defense, the other military services, government agencies, and civilian industry for coordination on the technical aspects of his evaluation. Seldom has there ever been such a broad and open-end charter in military history.

The basic Board membership was thirteen general officers and five civilians.[3] Some 3,200 military personnel and 90 civilians participated in various phases of war games, equipment and troop testing. Whole units were diverted from overseas deployments to be included in the tests. For example, a Caribou company which had been planned for the Pacific was delayed specifically by the Secretary of the Army, with the approval of the Secretary of Defense, so that it might take part in the tests.

The Office of the Secretary of Defense instructed the Department of the Air Force to support the Howze Board. On 8 May the Deputy Secretary of Defense, Roswell L. Gilpatric, requested the U. S. Air Force to contribute to the efforts of the Board through the use of its air transport capability, especially as a means of achieving air lines of communication within a combat theater. He stated that the Army would request up to a squadron of C–130's in its exercise and tests and that this would be a good chance for the U. S. Air Force to "sell" its services and capabilities. After detailed requirements were formulated, Secretary Stahr contacted Air Force Secretary Eugene M. Zuckert to work out arrangements between Continental Army Command and the Air Force Tactical Air Command.

The most significant major activity of the Board throughout its deliberations was the investigation, testing and evaluation of the organizational and operational concepts of airmobility. The findings and evaluations of field tests, war games, operations research, and visits to overseas combat theaters provided support to the final Board report.

Much effort was devoted to field experimentation for the purpose of comparing a conventionally equipped force with one made airmobile by adding aircraft. The Board had for a test force one battle group and part-time use of two others plus the artillery and engineers of the 82d Airborne Division. In addition, the testing

[3] General Howze was not only President of the Board, but Chairman of the Steering and Review Committee as well. In addition to General Howze, seven other officers and six top-level civilians originally composed the Steering and Review Committee. These included Major General Ben Harrell, Major General William B. Rosson, Brigadier General John J. Lane, Brigadier General Edward L. Rowny, Brigadier General Delk M. Oden, Brigadier General Robert R. Williams, Colonel William M. Lynn, Jr., Dr. Jacob A. Stockfisch, Dr. Edwin W. Paxson, Eugene Vidal, Fred Wolcott, Frank A. Parker, and Edward H. Heinemann. Mr. Parker, General Rowny, and Colonel Lynn also served as chiefs of working committees. Other senior board members (eventually added to the Steering and Review Committee) were named working committee chiefs—Major General Clifton F. von Kann, Major General Norman H. Vissering, Brigadier General Frederic W. Boye, Jr., and Brigadier General Walter B. Richardson.

committee used 150 Army aircraft, both rotary and fixed-wing, for eleven weeks, augmented by aircraft support from the U. S. Air Force. About fifty officers designed, controlled, and evaluated the tests assisted by an advisory group of civilian scientists.

The Board conducted approximately forty tests ranging in magnitude from fairly elaborate live-fire exercises and three major week-long exercises against an assumed force of irregulars to auxiliary tests of new items of equipment. Results of the three largest tests indicated that Army aircraft would enhance combat effectiveness in both conventional and counter-guerrilla actions and that Army tasks could be accomplished by smaller forces in shorter campaigns.

The Howze Board Report

The final report of the Howze Board was submitted on 20 August 1962. The air assault division was the principal tactical innovation. As compared with about one hundred aircraft in the standard division, it would have 459. Airmobility would be achieved by extensive reduction in ground vehicles from 3,452 down to 1,100 which would also reduce the airlift requirement for strategic deployment. Despite the reduction in total number of vehicles the cost of the new division, based upon the initial investment and a five-year operating cost, was about half again the cost of an infantry division. Artillery consisted of only 105-mm howitzers and Little John rockets (airtransportable in the Chinook helicopter). Augmenting this greatly reduced firepower, the division would employ twenty-four armed Mohawks and 36 Huey helicopters armed with 2.75-inch rockets.

The aircraft of the division could lift one third of its assault elements at one time. Three brigade headquarters provided major tactical sub-divisions to which the fighting battalions and support elements would be assigned according to the nature of the task and the terrain. The brigades in turn could mix units assigned to them as could the battalions and to a lesser extent the subordinate units of the battalions. All the essential elements of combat power—maneuver forces, reconnaissance, firepower, communications, and service support—were present.

The Board also recommended the organization of an air cavalry combat brigade having 316 aircraft, 144 of which would be attack helicopters. Its function was the classical one for cavalry—to screen, reconnoiter, and wage delaying actions. Unlike the air assault division which was designed to join battle on the ground, all of the air cavalry combat brigade was to be airborne including its anti-tank capacity.

A distinct feature of the proposed air assault operations was the requirement for an increased use of support aircraft to carry supplies as far forward as the tactical situation would permit. The Board envisioned Caribous and Chinooks located in the forward Army area to deliver men, equipment, and supplies after Air Force aircraft had deposited them as far forward as possible.

The Board presented Mr. McNamara five alternative programs which would modernize the Army structure by replacing conventional forces by airmobile forces. Of the five programs the Board recommended its Alternative Three as the "most responsive to the requirement and most compatible with the Army's mission and overall structure." This program within six years would provide eleven Reorganization Objective Army Divisions, five air assault divisions, three air cavalry combat brigades, and five air transport brigades. It would also increase the mobility of other combat units and the effectiveness of their logistical support.

The Board emphasized the quantitative and qualitative improvements that would be necessary for Army aviation personnel programs. Its proposed alternative would require 8,900 aviators in 1963, growing to 20,600 in 1968. The Board foresaw an increased need for warrant officer pilots and recommended an officer to warrant officer ratio of one to one by the end of five years. It also recommended major changes in the officer career program to enhance their training, administration, and utilization.

In a letter prefacing the final report, General Howze commented that the job was in some respects incomplete and further testing was recommended. He said, "The foregoing does not indicate that I consider the Board's findings unvalidated or its judgment faulty. The time made available although not sufficient to prove all details of the Board's recommendations as respects organization, personnel, equipment, maintenance and doctrine, was quite sufficient to enable it, with conviction, to chart a course of action which will serve to increase markedly the combat and logistical efficiency of the Army." The Board was aware that its proposals were not infallible and that subsequent tests and developments could alter some elements of the proposed force structure. Therefore, the Board recommended a continuing program of field tests using the first units becoming operational under the activation schedule. It also recommended that progress on the Board's proposals be reviewed annually as a safeguard against any errors in details of organization and any slippage in implementing the program.

The single major conclusion reached by the Board was terse and emphatic. "The Board has only a single, general conclusion," stated General Howze. "Adoption by the Army of the airmobile concept—however imperfectly it may be described and justified in this report—is necessary and desirable. In some respects the transition is inevitable, just as was that from animal mobility to motor."

The Howze Board accomplished its mission in 90 days, from the original assignment to the finished report. In view of the enormous staffing task involved, and the sheer size of the analytical task, such alacrity has few parallels in staff work in or out of the military service.

The Office of the Secretary of Defense, having received what it had asked for, now turned the report over to its systems analysts in the Office of the Comptroller to study in minute detail all 3,500 pages of the main report. The review by the office of the Secretary of Defense Comptroller was to focus on certain procurement actions and to make certain recommendations of its own which had no relation to the Howze Board's basic recommendations.

In mid-September 1962 the Chief of Staff directed General Howze, Colonel Norton, and one of the editors of the Board Report to work in his office in preparing rebuttals for the various attacks that were coming from all directions. This "ten-day" TDY mission was to last almost until Christmas (with General Howze being replaced by General Williams when the former had to return to his command at Fort Bragg as commander of the Strategic Army Corps during the Cuban missile crisis).

Throughout the fall of 1962 it appeared, at times, that the work of the Howze Board was going to be studied to death and finally filed away for historians. The fact that it survived attacks by members of Congress, the Air Force, and conservative elements within the Army was a tribute both to the soundness of its basic conclusions and to the dedicated officers within the Army who believed that airmobility was the wave of the future.

CHAPTER II

The Early Years in Vietnam, 1961-1965

The Army of the Republic of Vietnam Becomes Airmobile

There is no precise method to divide the Vietnam War into convenient phases. However, from the standpoint of an airmobility study, one can consider the first phase as a learning period—a time when U. S. Army pilots were teaching Army of the Republic of Vietnam commanders and soldiers how to effectively employ helicopter tactics, while at the same time the pilots were learning by experience, trial and error. As more and more helicopters became available, we built additional aviation units to help the Vietnamese Army become as mobile as the enemy.

This second phase of the war was characterized by battalion-size air assaults of selected Vietnamese units, including the paratroopers, the rangers, and the regular infantry. It was the success of this phase that forced the enemy to increase his effort in South Vietnam. This proved to be something that the North Vietnamese Army was quite ready to do, and the improved capabilities of the Army of the Republic of Vietnam were matched step-by-step with increased resistance of the Viet Cong and North Vietnamese Army, as additional units and supplies poured down the Ho Chi Minh Trail complex and across the border. It was during this second phase that we made great improvements on our tactical employment of helicopters. It was also during this period that we created our own airmobile division, tested it, and concluded that in terms of ground tactics, airmobility was here to stay. We studied the variety of tactics used by the Vietnamese and their U. S. support in these airmobile operations of extended scope and we tried to apply everything we learned to the organization and training of our airmobile units.

It was also during this second phase that the Huey came into its own. The turbine engine helicopter with its great power, its reliability, and its smaller requirement for maintenance, was the technological turning point as far as airmobility is concerned. Actually, the key improvement of technology was the trio of the

Huey as a troop lift bird, the Chinook with its larger capacity for resupply and movement of artillery, and the fledgling attack helicopter—these three together allowed us to take a giant step forward at this time.

From the time of the first major commitment of helicopters to Vietnam on 11 December 1961 until the buildup of major U. S. forces in 1965, airmobility was—like diplomacy—confined to the art of the possible. The early helicopters were old and unsuited for this particular mission; the rules of engagement were set by the South Vietnamese Government—a Government that was continually changing in a series of coups and counter-coups; and the Viet Cong was undergoing a phenomenal period of growth.

The primary mission of U. S. military forces was to advise and assist the Republic of Vietnam Armed Forces to build a viable military structure to meet the needs of its national security. Because of the increasing scope and intensity of Viet Cong operations, this advisory task expanded very rapidly during this period. The Viet Cong regular force grew steadily from two to five regimental headquarters, the Viet Cong battalions doubled in the same period, and the quality and quantity of their weapons and equipment improved considerably. This build up necessitated the deployment of additional U. S. Army aviation units to support Government of Vietnam forces. From a single transportation battalion with three helicopter companies in early 1962, the U. S. Army developed an enormous operational and logistical support complex consisting of many battalions of helicopter companies, fixed-wing units, maintenance units, and special purpose organizations.

Enemy Reaction

It is interesting to note how the enemy viewed this increasing airmobility potential. The following is a translation of an extract from a North Vietnamese instruction pamphlet captured on 16 November 1962:

> It can be said that all the recent augmentations of forces that the USA has sent to the Diem government were primarily intended to strengthen the Diem rear area forces, increase their ability to pass information rapidly and the wide employment of helicopters in the movement of troops. Therefore if we can destroy or greatly reduce the enemy's heliborne capability we will, in essence, have destroyed the mobility necessary to the US raid tactics.
>
> Although we have succeeded in inflicting some loss on the enemy in his heliborne operations the enemy has in some places caused us

fairly heavy losses. We must therefore find means of coping with the enemy's helicopter tactics. Widespread efforts must be directed to combatting heliborne landings and shooting at helicopters. Following are the advantages which the enemy enjoys due to his employment of heliborne strike tactics:

1. Careful planning and preparations are possible together with complete mobility in an attack, support or relieving role.
2. Secrecy can be preserved and surprise strikes can be accomplished.
3. Landings can be effected deep into our rear areas with the capability to attack and withdraw rapidly.
4. An appropriate means of destroying our forces while they are still weak.

However, these tactics suffer the following disadvantages:

1. The population in our rear areas is on our side and will resist the enemy in every way.
2. Small forces are usually employed by the enemy in their deep strikes and if counterattacked may find it difficult to withdraw.
3. Heliborne operations require the latest information (old info may have lost its timeliness and new info must be checked for accuracy). If the time is taken to acquire confirmatory info then the situation may have changed rendering the info inaccurate.
4. The enemy's strike elements are usually unfamiliar with the terrain and can easily be surrounded and rapidly defeated.
5. The present available helicopters prevent the enemy from employing large forces (although this is only a temporary disadvantage it will take the enemy some time before he will be able to overcome it.)
6. The effectiveness of heliborne tactics is greatly reduced in forested and jungle covered mountain areas where a clear knowledge of the nature of the terrain cannot be discerned from the air, where landings are difficult and ambushes easily employed against the landings.
7. The disadvantages inherent in helicopters are difficult to overcome. If they are flown slow or low they are vulnerable to ground fire; every flying hour must be complemented by 3 hours ground maintenance; they cannot be flown for more than 70 hours in any 2 or 3 days (TN: Obvious typographical or technical error. 10 hours in 2 or 3 days seems appropriate); the helicopter consumes much fuel, carrying a full load of troops its fuel capacity is reduced and as a consequence its range is reduced, as a result the starting point for heliborne operations is usually near the objective and thus the enemy's element of surprise can be compromised. A landing right within our position is the most effective, but also subject to coming under our firepower while a landing outside of our position, though avoiding our firepower, loses the element of surprise.

It is obvious that the Viet Cong, in spite of all their polemics, had recognized that the advent of airmobility in Vietnam had changed the name of the game. True, the allies had much to learn about the employment of this new capability, but the war would never be the same.

Early Problems

It is important to recognize that the early airmobility efforts with the H-21 and Army of the Republic of Vietnam personnel represented the lowest order of airmobility . . . that is, simply transporting people from point "A" to point "B". This is analogous to the requisitions of French taxis in World War I. It provided better and faster transportation than walking. But, it lacked the essentials of unified command, specially trained personnel, organic firepower, and responsive reconnaissance.

The early Army aviators in 1962 spent a great deal of their time simply training the Vietnamese in the rudiments of getting in and out of a helicopter and conducting themselves properly in a landing zone. This was further complicated by fuzzy command relationships. The U. S. Army helicopter pilot belonged to Military Assistance Command, Vietnam, and was attached to the U. S. advisor on a one-time mission basis. The aircraft commander still had the authority to abort a mission that he did not feel safe; yet, he was not in the early planning phases and did not share the responsibility of the success or failure of the mission. Furthermore, he did not control the tactical air support or artillery fire nor the timing thereof. It is easy to understand the frustrations some of the early relationships generated.

Little by little, sheer necessity forced those in-country to make workable agreements on standing operating procedures. The "who-does-what-to-whom" on command and control, flight levels, and fire support were among the most important procedures agreed upon. The evolution of a command and control helicopter, carrying all the essential commanders and liaison officers plus the proper communications equipment, soon came about. Armed helicopters would soon be part and parcel of every transport column—and the polywog shape of the Huey would soon become the universally recognized silhouette in Southeast Asia.[1]

---

[1] The transition to the Huey had its beginning in January 1955 when a design competition was held to select the new standard Army utility helicopter. The first model, designated the XH-40, flew in 1956 and the UH-1B model did not begin its user evaluation testing until 28 November 1960. By June of 1963, most of the light helicopter companies had phased in the new UH-1B helicopters and had transitioned their aviators, maintenance personnel, and crew chiefs to the new aircraft. But, it would not be until 27 June 1964 that the last tired old CH-21 was formally retired by General Oden, Acting Commanding General, U. S. Army Support Command, Vietnam. These times tend to give some appreciation of the long lead time necessary to develop and prove a new helicopter.

The Armed Helicopter in Vietnam

Early attempts had been made to arm the CH–21 with a light machine gun at the door, but this fire was relatively ineffective. To better meet this requirement, the Army formed the Utility Tactical Transport Helicopter Company and deployed it to Vietnam in mid-1962. This company was equipped with UH–1 helicopters armed with .30-caliber machine guns and 2.75-inch rocket launchers and was designed to provide protective fires for the CH–21 transport helicopters. Much of the tactical doctrine for armed helicopter employment evolved during this period including the techniques for protective fire preparation of landing zones prior to and during a helicopter assault. The Utility Tactical Transport Helicopter Company was redesignated the 68th Aviation Company and later the 197th Airmobile Company. Its early history was studied intensely by a special group known as the Army Concept Team in Vietnam which was established in Saigon on 6 November 1962.

General Rowny, after duty with the Howze Board, had been designated to form the Army Concept Team in Vietnam by the Chief of Staff of the Army with the mission to evaluate new methods of countering insurgency in actual combat. The Army Concept Team in Vietnam had a variety of projects to include rotary and fixed-wing aircraft, communications, armored personnel carriers, logistics, and civic action tests. Fortunately for the purpose of this study, they were able to document the early attempts at airmobility in a special and objective way.

One of the most important reports made by General Rowny's Army Concept Team in Vietnam evaluators was their analysis of the effectiveness of the armed helicopter company during the period 16 October 1962 through 15 March 1963. This test unit had a dual mission in that it was actually fighting a real war and providing armed protection for the transport helicopters, while, at the same time, it provided data to the evaluators (who sometimes were heavily engaged in combat themselves).

The first element of fifteen armed Hueys was deployed to Vietnam in September 1962. To assure proper employment, General Rowny hammered out a modus operandi with Military Assistance Command, Vietnam, on 29 September 1962, which provided a framework for the forthcoming test. The terms of reference provided that the test activities must not have an unacceptable impact on military operations. Therefore, testing was undertaken only in conjunction with actual operations, and in no case was the test unit required to engage in activities designed solely for test purposes.

More fundamental limitations were the rules of engagement for U.S. Army armed helicopters, which precluded testing of any tactical concepts involving "offensive" employment. Under these rules, the armed helicopters could deliver fire only after they or the escorted transport helicopters had been fired upon. In late February 1963 the rules were modified to permit the armed helicopters to initiate fire against clearly identified insurgents who threatened their safety or the safety of escorted transport helicopters.

Initially, the fifteen UH–1A helicopters were armed with locally fabricated weapons systems consisting of two .30 caliber machine guns and sixteen 2.75-inch rockets. In November 1962 the unit was augmented with eleven UH–1B helicopters. The "B's" were equipped with factory installed weapons systems of four M–60 machine guns per aircraft and locally fabricated clusters of eight 2.75-inch rockets. The "B" model eventually replaced most of the less powerful "A's."

The provisional Utility Tactical Transport Helicopter company was based at Tan Son Nhut Airport on the outskirts of Saigon and was under the direct operational control of Military Assistance Command, Vietnam. From this base, it supported transport operations of the 57th, 33d, and 93d Light Helicopter Companies, all equipped with CH–21 aircraft. In the latter part of the test period, one platoon of helicopters was sent to escort the Marine H–34 squadron operating in I Corps.

The plan of test for this company called for the evaluation of the armed helicopter in the "escort" role. Although "escort" was not defined, actual experience determined that the escort role broke down into an *enroute phase,* that was generally flown at a relatively safe altitude, the *approach phase,* where the heliborne force usually descended to nap-of-the-earth heights several kilometers away from the landing zone, and the *landing zone phase.* It was in the landing zone phase that the armed helicopter proved most valuable.

Prior to the advent of the escort by the Utility Tactical Transport Helicopter company, transport helicopters on the "dangerous" combat support missions were being hit at a rate of .011 hits per flying hour. For similar missions escorted by the Utility Tactical Transport Helicopter company, the rate declined to .0074. During this same period of time, the hit rate for all other flying done by Army helicopters rose from .0011 to .0024. In other words, the Viet Cong effectiveness against unescorted aircraft doubled while the efficacy of their fire against escorted aircraft dropped by 25 percent. Consequently, it was concluded that the suppressive fires delivered by armed escort helicopters were highly effective in reducing the amount and accuracy of enemy fires placed on transport helicopters. The response of the transport helicopter pilots to this added protection was clearly enthusiastic. Even the skeptical Marine pilots began to rely heavily on the Army armed helicopters.

Although the safety of the transport helicopters was the primary consideration in escort operations, the vulnerability of the escorts themselves was a matter of concern. The escorts preceded the transports into the landing zone and remained until all transports had departed. When the landing zone was small and the transport force used an extended formation, the period of exposure for armed escorts was unnecessarily long. The escorts generally flew at no more than 100 or 200 feet above the ground and were well within the zone of maximum vulnerability from small arms.

The Utility Tactical Transport Helicopter company flew 1,779 combat support hours from 16 October 1962 through 15 March 1963. Most of the operations were conducted in the III and IV Corps Tactical Zones. Suppressive fire delivered by the escort helicopters accounted for an estimated 246 Viet Cong casualties. During this period, eleven armed helicopters were hit by hostile fire. While no armed helicopter was shot down, one UH-1B was seriously damaged as a result of ground fire. It appeared that the vulnerability of armed helicopters was well within acceptable risk limits.

The first official Air Force recognition of the role of armed helicopters, and their first attempt to regulate their use in Vietnam, appeared in a document dated 27 December 1962, entitled "Helicopter Escort," and signed by Brigadier General Rollen H. Anthis, Commander of the Second Air Division. Generally, the armed helicopter was limited to one minute of fire before the transport helicopters landed and one minute after the last departed.

> . . . The fighter aircraft will have primary responsibility for the security of the entire helicopter formation during the escort phase until the formation commander makes the one minute warning call prior to landing. At this time the responsibility will shift to the armed helicopters and they will precede the transport helicopters to the landing zone to draw fire or engage targets. Armed helicopters may be designated to attack specific targets enroute when their assistance is requested by the fighter aircraft.

This arrangement proved unworkable and the Army continually sought more flexible guidelines. The Air Force consistently maintained the helicopter must be limited to strictly defensive fires and they were quick to report to Military Assistance Command, Vietnam, of an incident where the Army in their opinion usurped the role of close support. For example, on 26 July 1963, the Second Air Division wrote a letter to General Paul D. Harkins stating that, in three recent incidents, Army helicopters were used "offensively." Further, the Air Division inferred that: ". . . Employment of the HU-1B's in the above incidents appears to be an expansion of your directive for the use of armed helicopters to include interdiction and close air support missions. It also appears that air requests are being stopped prior to arrival at the ASOC's, with subsequent substitution of U. S. Army aircraft from Corps Advisor and Division Advisor resources."

Major General Richard G. Weede, USMC, Chief of Staff, Military Assistance Command, Vietnam, replied:

> . . . All incidents mentioned occurred in the immediate battle areas of the ground units. These are not subjects for ASOC/TOC coordination but rather matters for the ground commander to handle as he deems appropriate. Direct support aviation is controlled by the ground element commander and requires no supervision or control by a tactical air control system far removed from the ground battle.
>
> U. S. Army aviation when employed in a combat support role is normally under the operational control of the Corps Senior Advisors. Therefore, incidents of the type reported are of mutual concern of the ARVN Corps Commander and his U. S. Advisor. Request future incidents be referred to Corps Senior Advisors by your Air Liaison Officers for resolution at the operating level.

By mid-1963 the 1st Platoon of the Utility Tactical Transport Helicopter Company which worked with the Marine H-34's in the I Corps sector had become adopted by their comrades-in-arms as an integral part of their operations and few, if any, H-34 pilots elected to fly without the armed Hueys nearby. Procedures were developed whereby the armed Hueys picked up the fire support right after the fixed-wing fighter planes broke off their support for safety reasons, and that, in most cases, the last minute reconnaissance by the armed helicopters prevented the Marine H-34 from going into extremely hot ambushed landing zones.

Ironically during this same period, I found myself as Director of Army Aviation testifying in front of Congressional committees on the role and absolute necessity of more and better armed helicopters. The Commandant of the Marine Corps and various Air Force officers gave negative testimony on the merits of this system. Fortunately, the Army viewpoint prevailed.

After the tests of the armed helicopter and the Utility Tactical Transport Helicopter company were concluded and more Hueys became available for replacing the CH–21, the Department of the Army decided to convert the helicopter companies within Vietnam to a new airmobile company Table of Organization and Equipment. Each of the airmobile companies was organized into one armed platoon consisting of eight UH–1B helicopters with installed weapons systems and two transport platoons with eight UH–1B helicopters each. The Utility Tactical Transport helicopter company remained in-country and supported the Marines and U. S. Army units without organic armed helicopters.

The early tests with the Utility Tactical Transport helicopter company indicated that a platoon from five to seven armed helicopters could protect a transport helicopter force of from twenty to twenty-five aircraft. The organization of the new airmobile company was a compromise between the requirement to provide organic arms support and the requirement to lift troops and cargo.

The armament system brought the armed UH–1B up to its maximum gross weight thereby eliminating it from a troop or cargo-carrying role. In addition to the integrated machine gun and rocket systems, two door gunners were used on the armed helicopters. The door gunners provided additional fire when a threat was clearly identified, and they also performed the functions of clearing and reloading weapons between engagements and clearing some stoppages during engagements.

The UH–1B was not designed for an armed configuration and the weight of the armament system reduced the maneuverability of the aircraft and induced sufficient drag to lower the maximum speed to approximately 80 knots. As a consequence, the armed helicopters could not overtake the airmobile force if they left the formation to attack targets enroute. The early armed UH–1B's did an outstanding job in proving the concept of the armed helicopters, but they also pointed out many deficiencies that the Army would correct in later versions.

Techniques of Using the Armed Helicopter

Some basic operating procedures became standard. When a target was identified, the escort leader determined whether it could be attacked under the rules of engagement and, if so, directed the engagement. Where possible, targets were engaged at the maximum range of the weapon. This usually consisted of a continuous burst of machine gun fire throughout each firing run. The flight pattern was so planned that when the lead aircraft completed its firing run the next aircraft was in position to engage the target. This tactic placed continuous fire on the target until it was neutralized. Rocket fire was used when necessary to reinforce the fires of the machine guns.

The techniques of fire for all weapons systems were based on maneuvering the helicopter and manipulating the weapons during the firing run. The flexibility of the armed helicopter's weapons systems allowed targets to be engaged without a requirement to maintain the armed helicopter on an aircraft-to-target line. This feature was exploited whenever possible to minimize helicopter exposure to the target engaged. When using rockets to attack targets obscured by trees or overhanging foliage, it was found that best results were obtained with a delayed fuse.

Fire of the armed escort helicopters during the enroute phase was controlled by the armed escort leader. The pre-mission briefings identified the probable insurgent areas and prescribed fire control procedures to be used by both the armed escorts and transports enroute and during the landing phase. In the landing zone, before friendly troops were on the ground, armed escort fire was controlled by the escort leader. After the ground force arrived in the landing zone, the ground commander often marked and identified insurgent targets so that suppressive fire could be quickly and accurately placed on them. On medical evacuation missions and other missions of this type, when the ground unit was in control of the landing zone, the fire of the helicopters was controlled by the ground commander. He designated targets, marked friendly areas, and determined if escort fire could be used without endangering friendly forces. In mountain and jungle terrain, where targets were obscured, the suppressive fire had to be accurately controlled and fire discipline maintained in order to avoid hitting friendly troops and to prevent needless expenditure of ammunition.

The armed escort was under operational control of the transport commander during the enroute and landing phases of airmobile operations. If the armed escort remained in support of the troops in the landing zone, the ground commander assumed operational control.

During this period of time opinion varied on the necessity for door gunners on transport helicopters. Generally speaking, those not close to the action favored elimination of the door gunner for the additional weight and space, while the transport helicopter pilots favored the retention of the door gunner without exception.

Troop Helicopters Pick Up a Rifle Company From the Field

In mid-1963 the question of camouflaging the helicopters arose. Both camouflaged and uncamouflaged helicopters were used by U. S. Army units. Some pilots reported that the camouflaged helicopters were more difficult to see especially in mountainous terrain and a pilot spent an undue amount of his time keeping track of his other ships. Other pilots and evaluators expressed the opinion that the white Army markings and the yellow rings on the tails of the uncamouflaged helicopters were of value to the enemy gunners for the purpose of aiming and tracking. By and large, it was determined that the advantages of camouflage outweighed the disadvantages.

Tactical Troop Transport

When the UH-1B transport helicopter was first introduced in Vietnam, it usually carried ten combat-equipped Vietnamese soldiers and at times as many as eleven. An investigation determined that the average helicopter was grossly overloaded with this many soldiers. A combat-equipped Vietnamese soldier averaged 167 pounds. When ten personnel were loaded into the Huey with a full fuel load, a U. S. Army crew of four, armor plate, a tool box, a container of water, a case of emergency rations, weapons, armored vests for the crew, the Huey grossed 8,700 pounds, or 2,100 pounds over normal gross weight and 200 pounds over the maximum operational weight. Not only that, the center of gravity had shifted beyond safe limits. As a consequence, the standard procedure was to limit the UH–1B to eight combat troops except in grave emergencies. It was also directed that the armed helicopters would carry no more than the normal complement of five personnel and armament with the basic load.

It was the company commander's responsibility to insure that the helicopters were not overloaded and that they remained within maximum allowable weight and center of gravity limitations. However, because of operating requirements, the pilot was delegated this responsibility and would estimate the load to be lifted, supervise loading and tie-down, and insure that the aircraft was safe for flight. To verify the estimated safety factor, the helicopter was brought to a hover to check available cycling range and power prior to departure.

Methodology of the Early Air Assaults

During 1963 the single most important factor in the development of tactics, techniques, and procedures for airmobile units in the Republic of Vietnam was the lack of significant enemy air defense capabilities, either ground or air. The ground-based threat was essentially hand-held small arms and automatic weapons fire. On rare occasions caliber .50 or 12.7-mm machine gun fire was encountered. The lack of heavy enemy air defense had much to do with the selection of flight altitudes. During this time frame, most flights were made at 1500 feet or higher to reduce the chances of being hit by ground fire. Contour flying was rarely performed. The Viet Cong continued to ambush landing zones, especially in mountains or mangrove areas where there was a very limited number of landing sites. On occasion, they would mine the area or drive stakes to prohibit landing. Most of the resistance, not surprisingly, was in the critical landing phase of an air assault mission.

The planning for an airmobile assault had evolved rather rapidly from the haphazard coordination witnessed in early 1962. Usually, this planning was initiated when the aviation battalion received a mission request from the Corps Tactical Operations Center. Missions were then assigned to the aviation companies by the aviation battalion commander. The companies usually received daily mission requirements by 1800 hours on the previous evening. However, standing operating procedures were such that a mission could be initiated in less than an hour if necessary.

If time permitted, an aerial reconnaissance was conducted by the airmobile company commanding officer, a representative of the aviation battalion, and a representative from the supported unit. During the reconnaissance, approach and departure routes were selected, condition and size of the landing zone were noted, and flight formations, check points, and altitudes to be flown were determined. Reconnaissance air traffic over the intended landing zone was closely controlled in order to achieve the element of surprise.

Various companies used different helicopter formations enroute to a combat assault. A major consideration in the selection of a formation was the size and shape of a landing zone and the company commander's requirements for disembarking his troops after landing. A modified trail formation was sometimes used when an uninterrupted flow of troops into a small landing area was desired. The formation most frequently used was the "V". This formation proved to be versatile, easy to control, and permitted landing of the flight in a minimum of time without bunching. Helicopters normally flew about 45 degrees to the side and rear of the lead ship and high enough to be out of the rotor wash. Armed helicopters operated at the same altitude as the escorted force. A reconnaissance element of two or four armed helicopters preceded the transports by one to five minutes while the remaining escorts normally flanked the transports in a trail formation. If additional armed ships were available, they were positioned in the rear of the transports to engage targets under the flight.

The helicopter companies always attempted to plan return routes that were different from the approach routes. For subsequent lifts routes were varied slightly to avoid flying over a given area more than once. If one landing zone was used several times, the final approach and entry for each lift was varied if at all possible. In mountainous areas and in some jungle areas, it was not possible to vary the route of approach. Consequently, every attempt was made to land the troops in the shortest possible time to minimize danger.

Any deviation from the original air movement or landing plan was co-ordinated with the commander of the air-lifted force prior to execution. If the landing zone was heavily defended the mission commander could notify the commander of the air-lifted force and proceed to a pre-planned alternate landing zone.

The critical approach phase was initiated by all transport helicopters at the same time in an attempt to place all aircraft on the ground simultaneously. This was difficult to accomplish because of the stepped altitude of the formation, the rotor wash encountered during descent, and the difficulty in finding a suitable touchdown spot for each ship. However, it remained a goal. The terrain in the landing zone sometimes slowed the disembarking of troops. In the Delta, water was sometimes chest deep and the ship had to be held with the skids just under the water level or had to maintain a low hover. In jungle areas, grass ten to twelve feet high was often encountered. From the moment the first helicopter touched down until the last ship lifted off, two minutes were considered average unloading time for a twelve-ship formation. This two minutes seems an eternity when one is expecting enemy fire any second.

To lessen the possibility of fire being concentrated on a single ship, all helicopters attempted to depart at the same time. Direction of take off was varied for subsequent flights whenever possible. The armed escorts used the same tactics on the return route with the exception that the ships originally used for the reconnaissance were the last to leave the landing zone and consequently usually brought up the rear. If another lift was required, the formation returned to the loading area for troops and, if necessary, the aircraft were refueled and rearmed.

Perhaps the foregoing planning considerations sound trite to today's informed reader. But one must bear in mind that to the airmobile commander of 1963 they represented the distillations of many hard-learned lessons and the planning procedures were quite sophisticated for the situation. Even then they were being refined.

The Eagle Flight

In an effort to reduce the planning time required for executing an air assault mission, some of the earlier helicopter units developed a task force called an "EAGLE FLIGHT". An "EAGLE FLIGHT" was defined by Headquarters U. S. Military Assistance Command, Vietnam, as "a tactical concept involving the employment of a small, self-contained, highly-trained heliborne force. Tactical planning emphasizes the use of this force to locate and engage the enemy or to pursue and attack an enemy fleeing a larger friendly force. As an airmobile force, 'EAGLE' is also prepared to engage an enemy force located or fixed by other friendly forces. The inherent flexibility of the 'EAGLE FLIGHT' as a force ready for immediate commitment either alone or in conjunction with other forces is its most significant factor."

A typical EAGLE FLIGHT would consist of the following: one armed Huey would serve as the command and control ship and would have the U.S. Army aviation commander and the Army of the Republic of Vietnam troop commander aboard; seven unarmed Hueys were used to transport the combat elements; five armed Hueys gave the fire support and escort to the troop-carrying helicopters; and, one Huey was usually designated as a medical evacuation ship.

The EAGLE FLIGHTS were usually on a standby basis or sometimes even airborne searching for their own targets. Not only were these EAGLE FLIGHTS immediately available for those missions which required a minimum of planning, but they also provided the basis for larger operations. Several EAGLE FLIGHTS were sometimes used against targets that, when developed, proved too large for a single unit.

By November 1964, all helicopter companies in South Vietnam had organized their own EAGLE FLIGHTS and each company maintained at least one flight in an alert status on a continuing basis. The Vietnamese troop commanders were particularly enthusiastic about these operations for they provided a very close working relationship between the air and ground elements and a special esprit was built from the day-to-day operations.

Simply stated, the EAGLE FLIGHT was a microcosm of the large airmobile assaults that were destined to take place later. It had all the attributes of a true airmobile force with its self-contained reconnaissance and surveillance ability, firepower, and Infantry. Above all, these early EAGLE FLIGHTS were able to capitalize on the element of surprise which so often was lost in the detailed planning cycle with Army of the Republic of Vietnam forces.

The Growing Aircraft Inventory

At the beginning of 1964 the United States had 388 aircraft in Vietnam including 248 helicopters, too few to accommodate the expanding advisory effort and increasing Vietnamese Army operations. By the end of September 1964, there was in South Vietnam a total of 406 Army aviation aircraft supporting the Army of the Republic of Vietnam as follows:

O–1	(Birddog)	53
U–6A	(Beaver)	20
U–8	(Seminole)	9

OV-1	(Mohawk)	6
CV-2	(Caribou)	32
U-1	(Otter)	27
UH-1	(Iroquois)	250
CH-37	(Cargo Helicopter)	9

To support this effort, a total of 3,755 Army aviation personnel were provided consisting of 780 officers and 2,975 enlisted personnel. This made it possible to place a U.S. Army aviation company or a U.S. Marine Corps aviation squadron in support of each Vietnamese Army division with additional aviation supporting each Corps.

As 1964 came to a close, the U.S. Army support in Vietnam consisted of these major organizations: the 13th Aviation Battalion at Can Tho supported the IV Corps area with three Huey companies and one Birddog platoon; the 145th Aviation Battalion at Saigon supported the III Corps area with two Huey companies and one armed helicopter company; the 52d Aviation Battalion at Pleiku supported I and II Corps areas with two Huey companies, one Birddog platoon and one Caribou platoon; the 14th Aviation Battalion at Nha Trang supported all of South Vietnam and Thailand with one direct support maintenance company, one fixed-wing Otter company, and one Birddog platoon; the 765th Transportation Battalion at Vung Tau supported all four aviation battalions with two direct support maintenance companies, one general support maintenance company, one Caribou company, and one Special Warfare aviation detachment of Mohawks; and, finally, the Special Forces Group was supported with miscellaneous Army aircraft.

The Mohawk in Vietnam

Since I discussed earlier the controversial aspects of the Mohawk in its relation to the Army-Air Force roles and missions issues, it is important to review its actual performance in Vietnam in 1962. The six armed Mohawks did a magnificent job and many Army advisors pleaded for more of the same. The missions performed by these aircraft during this period were somewhat tangential to the story of airmobility in the Army. However, this does not take anything away from the individual accomplishments of the Mohawk pilots.

After a storm of controversy in the Pentagon, the 23d Special Warfare Aviation Detachment was deployed to Vietnam in Septemper 1962 for the purpose of providing air surveillance in support of Republic of Vietnam forces. In addition, they were to serve as a test unit for operational evaluation conducted by the Army Concept Team in Vietnam. The 23d Special Warfare Aviation Detachment (Surveillance) was organized in July 1962 as a prototype armed aerial surveillance unit using the OV-1 Mohawk aircraft. Besides its headquarters and photo processing section, there were three flight teams, each consisting of two armed Mohawks, four pilots, and seven enlisted maintenance and armament specialists.

When they were deployed to Vietnam their rules of employment specified that: on all operational flights a Vietnamese observer would be aboard; that the aircraft would be armed with .50 caliber weapons only; and, that this armament would be used only when required to defend against a hostile attack. The 23d Special Warfare Aviation Detachment was assigned to Support Group, Vietnam for administration and logistical support. Operational control remained with Commander, Military Assistance Command, Vietnam, who decided to place the Mohawks in support of II Army of the Republic of Vietnam Corps. From 16 October to 22 November the entire unit was stationed at Nha Trang supporting the 9th Division and the Railway Security Agency. On 23 November one flight team was moved to Qui Nhon, 100 miles north of Nha Trang, to be closer to the 9th Division.

The plan of test for the 23d Special Warfare Aviation Detachment called for systematic surveillance of a sector that would insure an even distribution of effort over a selected tactical area. However, it soon became abundantly clear that the supported units were generating so many missions for specific reconnaissance sorties within their tactical areas that no other missions could be flown. In each area, an element of the 23d Special Warfare Aviation Detachment was employed under the direction of the Army of the Republic of Vietnam commander, through his U. S. advisor, as an integral part of the total available force. Because of the many variables involved, the specific contribution of the Mohawk to the reduction of Viet Cong incidents could not be quantified. But, there was no question that the 23d Special Warfare Aviation Detachment inhibited and restricted Viet Cong activity.

Visual and photographic reconnaissance by this twin-turbine airplane produced a wealth of intelligence for supported units. Hundreds of structures, most of them camouflaged, were detected in Viet Cong base areas. Likewise, hundreds of people were sighted in suspect areas and, because of the detailed familiarity of Mohawk crews with the local situation and activity patterns, some of the people sighted could be positively identified as insurgents. One of

MOHAWK TAKING OFF

the unique advantages of the Mohawk in reconnaissance was its speed to noise relationship which allowed the aircraft to get within observation distance of people on the ground without alerting them to its presence. In one division, artillery fires directed from the air were nearly tripled by the activities of Mohawk observers. Supported units quickly developed air to ground procedures to exploit the capabilities of the immediately available Special Warfare Aviation Detachment aircraft.

From 16 October 1962 to 15 March 1963 the 23d Special Warfare Aviation Detachment flew more than 2,000 hours in the performance of 785 combat support missions. It had delivered defensive fire 27 times and had lost two aircraft. The cause of the loss of these two aircraft was never determined.

Many of the U. S. Army advisors wrote glowing reports of the Mohawk pilots and urged Military Assistance Command, Vietnam, to relax the rules of engagement. For example, on 23 March 1963 the Senior advisor of the II Corps Tactical Zone, Colonel Hal D. McCown, wrote:

> . . . Two Mohawk aircraft are constantly based at Quang Ngai airfield for close and immediate support of the 25th Infantry Division. Results of Mohawk operations there thus far include the following:
>
> a. Rapid production of low level aerial photographs of VC troop

dispositions and installations as well as photographic coverage of prospective operational areas of the 25th Infantry Division.

b. A large volume of combat intelligence has been produced by visual sightings of troops in the open including weapon emplacements.

c. A bonus effect has been obtained from the defensive machine gun fire put down by the Mohawks when fired upon. Some VC casualties have been observed from these fires.

d. During the period 15 February to 8 March, artillery adjustment by Mohawk observers on VC troops sighted accounted for an estimated 65 enemy casualties. Combining these casualties with those noted in subparagraph c above, it is concluded that the Mohawk has either directly or indirectly caused more VC casualties during this period than all other military forces in Quang Ngai, including RVNAF air strikes.

Furthermore, I consider that the attack by the well equipped Q 95 VC Battalion in the Bato area on 3 March was relatively impotent. This was most probably due to actions by Mohawk aircraft on the days immediately preceeding the attack. These aircraft were responsible for sighting and adjusting fire on large groups of armed VC within a few thousand meters of the scene of the attack.

Despite this fine performance, the full potential of the Mohawk aircraft cannot be realized because of the present test restrictions. I refer to the limitation of armament to the .50 cal machine guns.

The Mohawk aircraft consistently locate remunerative targets which are beyond the range of friendly artillery. The majority of these targets are small bodies of troops in the open. In counter-insurgency operations this is the type target most likely to be encountered. These are fleeting targets, and unless immediately engaged, will disperse and disappear. They are long gone before a friendly air strike can be mounted.

On the several occasions that the Mohawk has encountered and returned ground fire, the aircraft limitation to .50 cal machine guns has considerably curtailed the effectiveness of this defensive response. It is apparent that the addition of rockets, napalm and small fragmentation bombs would have greatly increased the number of VC casualties inflicted during the past few weeks. Further, this vastly increased fire-power capability would have a profound impact on the VC ability to conduct daylight troop movements. This would seriously hinder their tactical initiative, especially in Quang Ngai Province. It is also believed that this increased defensive firepower would enhance the surveillance capability of the aircraft by allowing greater low level freedom of movement.

In another letter, dated 19 March 1963, the Senior Advisor of the 25th Army of the Republic of Vietnam Division, Colonel Paul A. Baldy, echoed many of Colonel McCown's remarks and added his own list of impressive data. He closed his letter with the following:

It is my considered opinion that the Mohawks' success can be attributed to their immediate responsiveness to the commanders. In a counter-insurgency operation such as we have in Vietnam with its fleeting and elusive targets, the immediate and direct control over an aircraft as swift, silent and well-equipped as the Mohawk, flown by pilots with intimate knowledge of the terrain and the current enemy situation, is an invaluable asset to a commander.

Similar letters of praise came from the 9th Infantry Division advisors and the senior railway security advisor. In the case of railway security, enemy attacks decreased by 50 percent after regular surveillance was begun by the 23d Special Warfare Aviation Detachment. Nevertheless, the career of a railway engineer in Vietnam was considered to be non-habit forming.

There is no doubt that the Mohawks filled a real intelligence need for the U. S. Army advisors in the field. For the first time the advisors found themselves with a responsive tool in the form of the Mohawk under their direct operational control to fill in the many gaps in their intelligence. Their requests for aerial surveillance and fighter strikes through the cumbersome channels to the Vietnamese Air Force no doubt increased their desire for more Mohawk support. As can be seen from the above, there was the inevitable trend to use the Mohawk in the fighter role, mainly because it was there and it had the hard points necessary to carry armament.

The Caribou in Vietnam

Having described earlier some of the controversial aspects that surrounded the procurement of the Caribou by the U. S. Army, I must now touch on its role in the early years in Vietnam. Unquestionably, the Caribou filled a serious gap in the United States inventory. Its short field characteristics, pay load, and light "footprint" made it an ideal aircraft in the counter-insurgency environment.

A not-well-known fact is that the Caribou airplane predated our major helicopter commitment to Vietnam by five months. On 21 August 1961, an experimental "Y model" Caribou landed at Saigon under the auspices of the Advanced Research Project Agency. The Advanced Research Project Agency had its own charter to test almost any device, technique, or prototype that it felt might have application in our counter-insurgency efforts. Consequently, the Agency was able to cut through the red tape of Service disagreements and expedite those tests it deemed appropriate.

The Advanced Research Project Agency Caribou flew up and down the length of Vietnam going into airstrips normally limited to much smaller aircraft such as the O–1 or U–6 Beaver. Initially it specialized in resupply of Special Forces outposts that were inaccessible by road and that normally were supplied by either air drop or an occasional helicopter. In December 1961, while this aircraft was stationed at Da Nang, the Caribou was requested to support the Army of the Republic of Vietnam 2d Engineer Battalion which was attempting to improve the primitive airfield at A Shau. In view of the rainy season, the Vietnamese Air Force did not consider it suitable for either O–1 or U–6 aircraft; and, because of the constant cloud cover, air drops from C–47's were impossible. The Caribou navigated through the valleys to A Shau and landed on the wet primitive strip for the first time. Resupply parachutes from previous air drops and troops due for rotation were the first load out. The troops crowded to board the aircraft since none of them believed that the Caribou would dare to come back. However, the Caribou did return many times bringing rice, fresh food, and pierced steel planking to improve the runway. The loads of succeeding trips were dictated by priority, the weather always posing a threat. Finally, a complete road grader was moved in, 5,500 pounds at a time.

On December 17th, this Caribou took the President of the Republic of Vietnam and his official retinue into A Shau. Shortly thereafter, on Christmas Day, it landed President Ngo Dinh Diem and 37 additional passengers at a remote Montagnard outpost at Mang Buk in the II Corps Tactical Zone. This strip, not quite 1,000 feet long, was at an altitude of 4,000 feet and hemmed in by mountains. The Caribou had made an enviable reputation in Vietnam long before the first complete company of aircraft arrived in Southeast Asia.

On 23 July 1962, the 1st Aviation Company (Fixed-wing Light Transport) was self-deployed from the Continental United States to Thailand. In December the company was moved to Vung Tau, Republic of Vietnam. A second Caribou company, the 61st Aviation Company, was also self-deployed from the Continental United States in July 1963 and based at Vung Tau first. It was not until early 1963 that Commander in Chief Pacific approved the proposed test plan for the Caribou company, but by that time the Caribous had been integrated into most of the daily planning at the corps level.

The Army Caribou pilots took justifiable pride in their proficiency at bringing this big twin-engine airplane into the shortest possible strip. Even the old Southeast Asia hands of "Air America" were impressed when this lumbering bird made its unbelievably slow, steep approach into some of the primative airstrips. The two following examples were typical.

Sixty-five miles southwest of Saigon and just one-half mile north of the Mekong River, the Cao Lanh airstrip resembled an aircraft carrier's deck about three feet out of water. It was typical of the Delta strips with one noteworthy exception—its runway was only 55 feet wide. Since the Caribou had a main gear track of approximately 26 feet, and since a five to ten knot cross wind usually prevailed at Cao Lanh, the pilots had to exercise extreme precise directional control prior to and immediately after touchdown. While maintaining sufficient airspeed to facilitate directional control, they had to constantly guard against coming in too fast on the short 1,300 foot runway. From May to September the shoulders of this airstrip were too wet and soft for the aircraft to leave the runway. Consequently, the Caribou pilot had to make a half turn at the end of the runway, then shut down the inside engine while the local Army of the Republic of Vietnam security force pushed the aircraft back for another start, and then complete his turn. Nevertheless, the Caribou flights served this strip three times a week.

Probably the shortest strip used consistently by the Caribou was Tra My in I Corps. Tra My was 830 feet long and lay in a valley floor requiring very steep descents and maximum performance climb-outs even for aerial delivery. At each end of the strip was an abrupt embankment. There was no room for a short landing or an overrun. The first landing was made by Captain Ephriam M. Goss and CWO P. Crossan. After three attempts, Captain Goss touched down twelve feet from the west embankment and came to a skidding halt eight feet from the east embankment. Needless to say, a hasty effort was made to lengthen the strip.

During the first six months of formal testing in early 1963, the Caribous flew 47,563 passengers and carried over 3,800 tons of cargo in over 9,000 sorties. Its availability rate was consistently high. It proved a boon to Army logistical support in Vietnam. General Rowny, who was in charge of the tests being run by the Army Concept Team in Vietnam, stated:

> The first Caribou tests there used the plane to support Green Berets located in inaccessible places. Later, all sorts of cargoes were lifted into all kinds of fields. Montagnards were resettled (complete with chickens and cattle); fuel was delivered to stranded helicopters; casualties, when they occurred in large numbers, were quickly lifted to hospitals; spare engines and parts, fresh water, food and other vital cargo were lifted. Soon there were more demands for the Caribou than we could meet. Added to its advantages of reliability and versatility was its availability. There were no long lead-times needed for processing requests; no tortuous channels to go through. If a commander felt he needed logistical

help he merely called his supporting Caribou detachment commander. Seldom was a request denied; almost never was he disappointed with the performance. The Caribou tests were highly successful—the light cargo plane seemed assured a place in the Army inventory.

Like the Mohawk, the history of the Caribou is somewhat tangential to the history of airmobility in the Army. It *did* perform a logistics role in a gross weight and performance category that was unique. The next plane in size was the C–123B which had a gross weight of 55,700 pounds compared to the Caribou's 28,500 pounds, and the single wheel load of the C–123 was almost double that of the Caribou. Consequently, the Caribou could operate repeatedly and routinely into airfields that were denied to the C–123 and later the C–130. For example, in 1963, the Caribou could operate into 77 percent of all airstrips in Vietnam while the C–123 was limited to just 11 percent of these airfields. The big question at this time was whether the Caribou should be employed under the corps senior advisors or whether it should be integrated into the Southeast Asia airlift system. Strong differences of opinion existed.

In 1964 the 61st Aviation Company introduced the low-level extraction system to Vietnam. In this technique the Caribou flew over a restricted drop area, at an altitude of about ten feet, and a drag chute attached to the load which was to be deployed pulled the load out of the ship through the rear door. Faster than the time it takes to tell, the load skidded across the ground until it came to a safe rest. Using this system, a narrow stretch of road was an acceptable drop zone. During my visit to Vietnam in August 1964, the only complaint I heard about the Caribou was that there weren't enough of them.

Other Army Aviation Units in Vietnam

At that time our fixed-wing aviation assets were centralized in Vietnam in the Aviation Support Battalion (Provisional) commanded by Lieutenant Colonel Robert J. Dillard. This battalion consisted of the 18th Aviation Company (U–1A Otters) for light transport, the 73d Aviation Company (O–1F Bird Dogs) for reconnaissance, the 61st Aviation Company (CV–2B Caribou) for heavy transport, and the 23d Special Warfare Aviation Detachment (JOV–1 Mohawk) for surveillance. Three transportation maintenance companies provided the necessary logistics support. The 18th Aviation Company with its Otters had been doing yeoman service in Vietnam since January 1962 and its pilots generally felt that their unique contribution to supplying infinitesimal airstrips with everything from pigs and chickens to ammunition had been overlooked in the press attention to more exotic units. The lumbering Otters flew on and on with little maintenance and little credit.

Another unit which is seldom mentioned in the Vietnam reports (probably because it was so often taken for granted) was the 73d Aviation Company flying the two-place Bird Dog of Korean vintage. That company had arrived in Saigon on 23 May 1962 and its thirty-two aircraft were spread in fifteen separate locations all the way from Hue in the north to Bac Lieu in the south. These O–1's were primarily oriented to the reconnaissance requirement for the Vietnamese advisors but also were utilized for artillery adjustment, target acquisition, command and control, message pickup, medical evacuation, radio relay, and resupply. By 1964 this unit had set up its own school to train the Vietnamese officers and aerial observers and they had recorded over 41,000 hours of flying time in their first fourteen months in Vietnam.

Increasing Viet Cong Threat

While I was in Saigon in the summer of 1964 I spent some time with Major General Oden, Commanding General of the U. S. Army Support Command in Vietnam. General Oden, who had been my predecessor as Director of Army Aviation, highlighted the increased intensity of combat in Vietnam and the concurrent increased risks being taken by our aviation units. It was apparent at this time that the most important aspect of U. S. support to the Army of the Republic of Vietnam was the capability embodied in our U. S. Army aviation organizations.

From the arrival of the first H–21's in December 1961 up to mid-1965, the U. S. Army had concentrated on developing airmobile operations in support of Army of the Republic of Vietnam forces in an ever-increasing scope. As the Allied tactics and techniques developed, so did the Viet Cong develop counter tactics and techniques. Consequently, there was always a need for innovation. Experience proved that any set pattern for any length of time was extremely dangerous since the Viet Cong were quick to capitalize on these patterns and strike at the weakest point.

Commander, Military Assistance Command, Vietnam, on 12 June 1965, indicated his concern of the number of helicopter assaults that had been ambushed by the Viet Cong in recent months and directed Brigadier General John Norton, Deputy Commander of U. S. Army Support Command, Vietnam, to study this problem.

General Norton directed each corps to make a special presentation on their helicopter operations at a critique on 5 and 6 July 1965. After the critique, separate study groups were formed to discuss in depth the enemy and the environment, airmobile operations from planning to execution, command and control and communications, and current and prospective material. The meeting served to summarize the problems of airmobility during this period and possible solutions for the future.

General Norton's group noted that the Viet Cong were increasing their efforts to counter friendly airpower with larger caliber weapons and many reports indicated the introduction of the 12.7-mm machine gun into South Vietnam. It was reasonable to expect heavier antiaircraft weapons up to 37-mm in size to be introduced within the next six months.

The Viet Cong had learned more than a bit about the method of operation of Free World Forces and could well determine probable landing zones and the number of troops which could be brought in by one lift. Their antiaircraft weapons were now being centrally controlled and coordinated to deny the use of the most desirable landing zones and thereby channelize the airmobile forces into landing zones chosen and covered by the Viet Cong. In the latter zones, the Viet Cong forces tried to maintain favorable odds of four to one to the amount of Army of the Republic of Vietnam forces available to react against them. Even in 1965 it was envisioned that soon airmobile operations would be opposed by division-size units.

Combat intelligence for airmobile operations was woefully inadequate due to a multitude of inadequacies in Army of the Republic of Vietnam intelligence combined with restrictions imposed in U.S.–Army of the Republic of Vietnam advisory relations. The increased activity of the enlarged Viet Cong forces made the staging areas of airmobile forces increasingly vulnerable to surprise attacks. Fuel and ammunition depots were prime targets.

Increasing Viet Cong interdiction of main ground supply routes and the resultant dependency on air-delivered supplies was imposing a severe burden on available air cargo capabilities. It became evident that special command control had to be vigorously applied to insure that critical air tonnage was not diluted by non-essential cargo.

General Norton's study group also highlighted one of the major problems of the early armed helicopters—the armed Huey because of its gross weight and additional drag was slower than the troop-carrying transport helicopters that it escorted. This meant that either the whole column slowed down or that the assault had to be timed so that the gun ships and troop ships rendezvoused just prior to landing. There was an immediate requirement for a faster armed escort helicopter which could maintain a speed of at least 150 knots. The urgency of this requirement was to force the Pentagon to re-examine its position on the introduction of an "interim" armed helicopter rather than wait for a totally new weapons system.[2]

In summary, in June of 1965, the U. S. Army found itself with a large commitment of airmobile resources supporting Republic of Vietnam Armed Forces with an organization that had grown somewhat like Topsy. Tactics and techniques had been generated by the necessity of the moment, procedures had been hammered out of necessity, and equipment had been borrowed and jury-rigged. Airmobility had obviously kept the South Vietnamese forces in being, but the Viet Cong had become increasingly more sophisticated and were reinforced with large numbers of North Vietnamese regular troops. The U. S. Army aviators and crew chiefs had done a heroic job in the face of countless difficulties, but it was obviously time to develop a new order of magnitude of airmobility and regain the offensive. The President of the United States decided to commit major numbers of U. S. ground combat troops to the action.

The option open to the President in 1965 to deploy a significant U. S. airmobile force was only possible because certain events had occurred in the preceding years. To understand airmobility in Vietnam we must examine the gradual development of the concept in the United States. Before we proceed with this examination, I want to interject one important reminder. The airmobile concept had its roots in the necessity to counter a sophisticated enemy in Europe. The Howze Board had only touched on the counter-guerrilla possibilities and the Air Assault tests, which we will discuss next, centered around two conventional opposing forces. Vietnam was in the background, to be sure, but only as one of many contingencies—not *the* contingency.

---

[2] A detailed discussion of the development of a "new" armed helicopter will be found on page 144 where we discuss the arrival of the Cobra in Vietnam. Here it is sufficient to note that by this time in 1965, the concept definition of an Advanced Aerial Fire Support System had evolved from a rather straight-forward compound helicopter to an ultra-sophisticated, all-weather, computerized weapons platform with a rigid rotor. It's my personal opinion that the original concept could have been developed and delivered in this time frame, if the requirement had not been unduly complicated by additional requirements that strained the state of the art, increased costs, and delayed delivery.

CHAPTER III

The Early Years in the United States, 1963-1965

The Air Assault Tests

Before we examine the details of the 11th Air Assault Division and the eventual deployment of the 1st Cavalry Division (Airmobile), it is important to remember that during this period the Aviation School, Transportation School, Signal School, and Aberdeen Proving Ground, as well as many other Army agencies, devoted their considerable talents and manpower to improving and supporting the concept of airmobility. Space and time do not permit detailing even a portion of this effort, but I would not want any omission here to leave the impression that the air assault tests were the sum and substance of the U. S. Army progress during this period.

The Howze Board had laid the foundations and suggested the means to finish the job that it had begun so well. Now it was up to the Army as a whole to turn the major Howze Board recommendation into fact.

To maintain the Howze Board momentum and to meet one of its major recommendations, Deputy Chief of Staff for Operations on 7 January 1963 issued the initial plan for the organization, training, and testing of an air assault division and an air transport brigade. Cadres of the test units were activated on 15 February at Fort Benning, Georgia. The test division was named the 11th Air Assault Division to revive the colors of that proud World War II airborne division. At first it was represented by just an infantry battalion, plus a few personnel in the division headquarters and the necessary combat support and logistical support elements. Concurrently, the 10th Air Transport Brigade was activated around a battalion-size unit. The overall strength of these initial test units was 191 officers, 187 warrant officers, and 2,645 enlisted men for a total of 3,023 personnel. Of the 154 aircraft provided, 125 were helicopters and 29 were fixed-wing. It was visualized that the units would be progressively expanded to full strength beginning in Fiscal Year 1964.

Within the Army staff there was considerable anxiety that the Army would be pushed into joint testing of these new organizations before they were sufficiently organized and trained. General Barksdale Hamlett, then Vice Chief of Staff, stated on 11 February 1963 that his chief concern at the moment was whether the Army would "be permitted to pursue an orderly program without being forced into premature joint testing." He believed that "within our airmobile concept the principal targets for the opposition appeared to be the Mohawk, the Caribou, and the armed helicopter together with our plans for the utilization of these aircraft." Later events were to prove that General Hamlett's concern was well founded.

Those who were assigned to the new test units throughout the year and a half of the formation and testing of the 11th Air Assault Division and the 10th Air Transport Brigade almost universally regarded this period as one of the high points of their lives. It was one of the few times in the Army up until that point in time that a group of officers and men have been pulled together with the job of developing and proving a concept with very little in the way of approved doctrine, systems, equipment, methods of operations, and any of the vast documentation and regulations which normally prescribe the formation of new military organizations. For example, at the time of the formation of the 11th Air Assault Division, it was actually against Army regulations for helicopters to fly in formation except under the most unusual circumstances. The infantryman had to adjust to new methods of entering into combat and new tactics and techniques of closing with the enemy. The artillery man had to provide his proven support with new airmobile artillery and aerial rocket artillery. The aviation elements had to broaden their training to include much work in the nap-of-the-earth, formation flying, night formations, jury rigging of weapons on Hueys and Mohawks, and forward area refueling. It was a time of innovation at all levels.

Brigadier General Harry W. O. Kinnard had been selected to lead the 11th Air Assault Division during this critical period. He in turn handpicked his key personnel and gave them the widest latitude in accomplishing their particular portion of the mission. Commanders at all levels were free to pursue vigorously any advancement of the airmobile concept as they saw it. At division headquarters, General Kinnard established "an idea center" to insure that any suggestion however bold or radical would receive careful and detailed consideration. Civilian industry was briefed on the test program and submitted a varied array of new equipment items for testing. Aviation weather minimums were relaxed to permit the launching of airmobile operations depending on the state of training and the skill of the flight crews. Particular attention was given to night operations.

Lieutenant Colonel John B. Stockton's 227th Assault Helicopter Battalion—the first in the Army—spent much of its time experimenting with methods of moving long distances through very low weather and improving their own lighting systems for tight formation flying at night. The Chinook battalion under Lieutenant Colonel Benjamin S. Silver, Jr. struggled with a newly developed machine, with the attending problems of maintenance and spare parts, to find new methods of moving artillery and key supplies. The 10th Air Transport Brigade, under Colonel Delbert L. Bristol, with a combination of Caribous and Chinooks devised the first workable air line of communications.

While the test units were being formed and organized, considerable thought was being given to the methodology that would have to be developed to objectively test such large organizations. There had been no tests of this type since the years just prior to World War II, when the Army tested its "new" triangular division concept. And even this test never approached the scope of the Air Assault Tests. There had been, of course, many large unit field exercises, but these took the form of training tests rather than concept tests. Furthermore, the evaluation of exercises of this type had been based solely on military judgment and opinion.[1]

Lieutenant General C. W. G. Rich, who was assigned the overall responsibility as test director on 1 August 1964, was given the mission to test a *concept,* and base the evaluation of this concept on scientific principles. He charged the Test Evaluation and Control Group, headed by General Williams, to establish a new methodology based on evaluation of the combat systems and how these systems interacted with each other. A scientific element, staffed by systems analysts from the Combat Operations Research Group, advised and assisted the evaluation group. In the final phases of evaluation there were 376 permanent members of the Test and Evaluation Group as well as 1,596 temporary personnel.

The main problem facing the Test and Evaluation Group was the difficulty of isolating *what* was being tested. Test units were developing procedures, detailed tactics, and techniques through

[1] Extensive tests of the "Pentomic Division" were made at Fort Benning in the mid-1950's, but they differed in magnitude and approach.

the period of observation. Thus, a system and its associated procedures which was observed one day would not necessarily agree with an observation of the "same" system the following day. To further compound this problem, the testing units had to train concurrently with the testing, so the test findings had to be analyzed as to whether they were the result of unit training problems or actual deficiencies in the system itself.

Throughout the early formation and training period of the 11th Air Assault test units, there was a continuous and intentional cross-feed of people, information, equipment, and ideas between what was going on in Vietnam and what was going on at Fort Benning. Members of the 11th Air Assault paid frequent visits to units in Vietnam for cross-fertilization of ideas, and many of the members of the 11th Air Assault in its latter stages were Vietnam returnees. In addition, the test division had the added mission of forming, training, and equipping a total of six airmobile companies that were sent to Vietnam during the testing period. The airmobile concept was not growing in isolation either in Southeast Asia or the United States.

It was not until three months before the final test, labeled AIR ASSAULT II, that either the division or the air transport brigade existed as essentially anything more than a strong cadre organization. Between 1 July 1964 and 1 September 1964 both units were levied for complete units with aircraft and personnel for shipment overseas. This requirement skimmed off the cream of highly qualified individuals from all units in the division. Temporary duty aircraft and aviators were brought into the testing units during the month of September and continued to join the testing units up to and during the final exercise which took place from 14 October to 12 November 1964.

AIR ASSAULT II involved some 35,000 personnel and covered over four million acres through the Carolinas. The exercise was a controlled field test where a test scenario listed by time certain actions or events that were required to occur. Maximum freedom of player action was permitted with sufficient control to allow adequate observation for those events which were necessary for evaluation. The 82d Airborne Division (augmented) acted as the aggressor force.

As fate would have it, the weather at the beginning of AIR ASSAULT II was incredibly bad. Hurricane Isabell, out in the Atlantic, had low flying scud and rain blanketing out the whole of the eastern seaboard. Ceilings ranged from 50 to 200 feet and visibility was sometimes less than one eighth of a mile. These conditions were worsened by gusty winds and low-hanging haze and fog. All the airlines on the eastern seacoast had ceased operating since turbulence above 1,000 feet was so severe that instrument flying was not feasible. Despite this, this operation jumped off only one hour behind schedule with the movement of one air assault infantry brigade by 120 helicopters over a distance of 100 nautical miles precisely on their objective. A year and a half of training had paid off.

For four weeks the test units maneuvered throughout the Carolinas in offensive, defensive, and retrograde movements. The umpires were hard-pressed to keep up with the tempo and the collection of "hard data," by being in the right spot at the right time, became their primary concern. So much was happening concurrently over such a large area that control of the enthusiastic units being tested was a constant problem. It was a tribute to both the tested units and the umpire personnel that the scenario remained recognizable. To compound the problem, hundreds of distinguished visitors wanted to witness this critical testing period.

It is interesting to compare the comments of the two major commanders of the units in AIR ASSAULT II, Major General Robert H. York, who commanded the aggressor force of the 82d Airborne Division, and General Kinnard, who commanded the 11th Air Assault Division and attached units.

General York said:

> Air assault operations as pioneered on Exercise AIR ASSAULT II have a dynamic potential. Seldom do we see a new military concept which can contribute so decisively throughout the entire spectrum of warfare. Certain air assault techniques used during Exercise AIR ASSAULT II would be unacceptably hazardous in actual combat. However, these deficiencies can be corrected and do not detract from the validity of the overall concept.

General Kinnard looked beyond the scenario of AIR ASSAULT II and said:

> Beyond what I believe to be its capabilities to perform roles normal to other divisions, I am even more impressed by what I feel is its ability to perform in unique ways beyond the abilities of other divisions. For example, in a low scale war, I believe it can exert control over a much wider area and with much more speed and flexibility and with much less concern for the problems of interdicted ground communications or of difficult terrain. In higher scales of war, I see in this division an unparalleled reserve or screening force capable of operating over very large frontages. By properly picking times, places, and methods, I believe it can also operate with devastating effect against the rear of the enemy. Faced with threat or use of nuclear weapons, I believe it

can widely disperse and yet, when required, quickly mass (even over irradiated ground, blown down forests or rubbled cities), strike an enemy, then disperse again.

On 1 December 1964 General Rich forwarded his interim report of the Air Assault test to the Commanding General, Combat Developments Command. He noted that this examination of a particular division slice had not set the lower or upper limits to the airmobility potential but it had provided clear indications of the possible and practical advantages to be gained. His recommendations were as follows:

> *First,* I urge that the two years of effort, the experience of the people on hand, and the equipment on hand not be lost by the dissipation, fragmentation, or dispersal of the tested units. *Second,* I strongly recommend that the 11th Air Assault Division or a division strength unit with the airmobility capability of the 11th AAD be included in the Army's force structure with a full parachute capability for its non-aviation elements; and the 10th Air Transport Brigade be retained intact and included in the Army's force structure. The significant question is not whether we can afford such organizations, but whether this nation, with its rapidly expanding population and ever-increasing GNP can afford *NOT* to have them. The tested organizations are prototypes, in being, of the most versatile forces that we can add to the United States Army. The *movement* capability of all divisions, including the 11th Air Assault Division has been enhanced by Air Force aircraft; however, the integration of Army aircraft into these tested units has provided the crucial *maneuver* capability of light mobile forces to close with and destroy the enemy. In combination with ROAD divisions and other standard Army organizatons, airmobile units offer a balance of mobility and an increased Army combat readiness on a theater scale that is applicable to the entire spectrum of warfare.

Though not part of the formal report, General Rich made one further observation on the fundamental concept of airmobility.

> I wish to distinguish between three fundamental levels of airmobility. First, an aviation unit can be given to a combat force on a temporary basis for a specific operation. This is equivalent to a corps truck company attached to a division for a one time move. Such an operation involves two separate staffs working out detailed plans to integrate the SOP's and techniques of two separately trained organizations. The second level is represented by the organic aviation in a ROAD Division. This approach benefits from unity of command, day-to-day training and intangibles such as esprit. But it is limited to a company lift capability; does not permit replacement of ground vehicles by aircraft; its equipment is not tailored to aircraft capabilities; and it could never represent the primary thrust of the division. At the third level a much greater gain is possible when the organization is specifically trained and equipped to exploit the continuing close tactical integra-

tion of heliborne lift as a primary means of maneuver, accompanied by readily available aerial fires and by highly responsive aerial reconnaissance and support systems. In my opinion, the combat power offered at these three levels rises on a geometric, rather than an arithmetic scale, and only at the third level do we find a new potential in the tempo of operations, in range over extended distances and in freedom from heretofore formidable terrain obstacles.

Joint Considerations

As early as 17 January 1963 the Joint Chiefs of Staff had envisioned testing the airmobile concept under the supervision of U. S. Strike Command. It was originally conceived that this would be a complete comparative evaluation of the Army and Air Force airmobility concepts to include division-size joint exercises. General Rosson was designated by General Paul D. Adams, Commanding General U. S. Strike Command, to form a Joint Test and Evaluation Task Force to plan and fulfill this requirement. As mentioned earlier, the Army was concerned that it would be pushed into such a test before it had fully organized and trained its own airmobile units. The Air Force, on the other hand, had been pushing for a joint test of its own concepts ever since the Disosway report on the Howze Board.

At the heart of the Air Force concept was the contention that within a joint force the Army Reorganization Objective Army division supported by Air Force tactical air offered a more practical and economical means for enhancing the mobility and combat effectiveness of Army units than did the Army air assault division. Selective tailoring of the Reorganization Objective Army division was seen as permitting varying degrees of airtransportability and combat capability ranging from a relatively light mobile force to one capable of sustained combat, all without recourse to specialized Army airmobile units. As visualized by the Air Force, neither Army fixed-wing aircraft or medium helicopters were required for tactical movement of troops or delivery of supplies. It was their contention that the Air Force C–130 could do the majority of the air transport mission while other Air Force aircraft provided reconnaissance and firepower.

On 5 March 1964 a decision by the Secretary of Defense produced far-reaching effects on the U. S. Strike Command test and evaluation effort. This decision allowed the Army to proceed with its unilateral tests during the latter part of calendar year 1964. Thereafter, the Army was to recommend to the Joint Chiefs of Staff what aspect, if any, warranted validation by joint testing.

The Joint Chiefs of Staff would then determine if there was a requirement for joint testing of the Army's concept during calendar year 1965.

Meanwhile Strike Command was left with the responsibility for conducting joint training and testing of the Air Force concept through division level. The 1st Infantry Division subsequently was made available for this purpose.

The Strike Command-sponsored joint test and evaluation exercise, GOLD FIRE I, was conducted in Missouri at about the same time that the Army was conducting its final test program with the air assault division. It soon became evident that the Air Force concept, rather than dealing in innovations, embraced improving streamlined sustained execution of a long-established concept; namely, tactical air support of ground forces. It proved that with overwhelming use of dedicated Air Force support, a standard Army division had increased potential. The joint exercise provided few surprises.

General Harold K. Johnson, Chief of Staff, United States Army, in discussing the differences between the tests of the 11th Air Assault Division and the test sponsored by Strike Command of an Army division supported by the Air Force, remarked: "I had the rare privilege of seeing the 11th Air Assault one week and the other concept at the early part of the following week, and I would make a comparison of perhaps a gazelle and an elephant. The two are not comparable. Each of them has its role to play, and it is important that we continue to pursue in this area where we have made such significant strides the gains that we already have."

At the last moment, General Rosson was told by the Chairman of the Joint Chiefs of Staff on 12 September 1964 to carry out an independent U. S. Strike Command evaluation of the Army's AIR ASSAULT II and on 6 October 1964 the guidance was modified to state that a comparative evaluation of GOLD FIRE I and AIR ASSAULT II would not be undertaken but rather directed a separate and independent evaluation of the Army exercise.

The interim final report on the tests of the 11th Air Assault which was submitted on 1 December 1964 essentially wrapped up twenty-one months of intensive, almost feverish, activity by thousands of highly specialized personnel. General Rich's final plea had been that this organization not be dispersed to the winds. It was now up to the Office of the Secretary of Defense to determine its future.

The Army had convinced itself, and a large body of people throughout the military establishment, that further large scale

CHART 1—1ST CAVALRY DIVISION (AIRMOBILE) ORGANIZATION

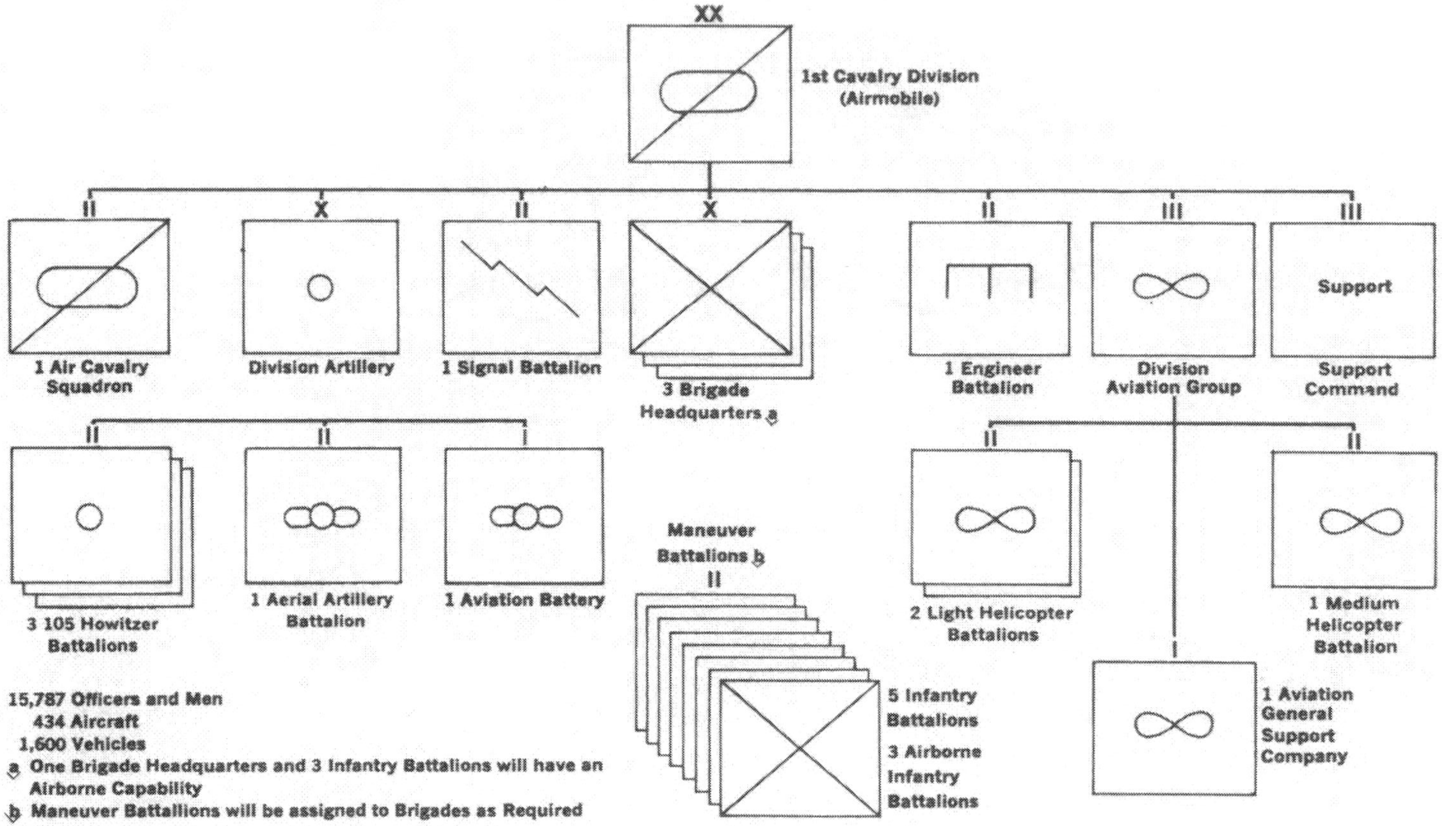

ADMINISTRATIVE DIVISIONS
SOUTH VIETNAM
Da Lat AUTONOMOUS MUNICIPALITY
0 25 50 75 100 MILES
0 25 50 75 100 KILOMETERS
QUANG TRI
Hue
THUA THIEN
Da Nang
I CORPS
QUANG NAM
QUANG TIN
QUANG NGAI
KONTUM
BINH DINH
Qui Nhon
PLEIKU
PHU BON
PHU YEN
DARLAC
KHANH HOA
Nha Trang
QUANG DUC
TUYEN DUC
Da Lat
NINH THUAN
Cam Ranh
PHUOC LONG
BINH LONG
TAY NINH
LAM DONG
BINH DUONG
LONG KHANH
BINH TUY
BINH THUAN
II CORPS
HAU NGHIA
BIEN HOA
SAIGON
CHAU DOC
KIEN PHONG
KIEN TUONG
LONG AN
GIA DINH
PHUOC TUY
AN GIANG
DINH TUONG
GO CONG
III CORPS
SA DEC
VINH LONG
My Tho
KIEN HOA
Vung Tau
Rach Gia
KIEN GIANG
Can Tho
PHONG DINH
CAPITAL SPECIAL ZONE
CHUONG THIEN
VINH BINH
BA XUYEN
IV CORPS
BAC LIEU
AN XUYEN

MAP 1

tests were unnecessary to make the critical decision as to whether to keep such an organization as the air assault division in being. Strike Command was unsatisfied with the unresolved joint operational problems associated with these new units, but no funds had been programmed for major tests during calendar year 1965. As thousands of pilots, mechanics and technicians returned to their home stations from temporary duty, the remainder of the test personnel finished writing annexes to the final test report, completing documentation of side tests of new equipment, and pulling long over-due maintenance on their tired aircraft.

On balance, it must be said that the continuation of active service of the 11th Air Assault hinged not only on the test results per se, but also the events in Vietnam which made the deployment of such a division extremely attractive. In the spring of 1965 many planners in the Pentagon were considering the deployment of this division as one of the possible options to counter a worsening situation—one in which the Viet Cong seemed likely to cut South Vietnam in half through the II Corps Tactical Zone.

Formation of the 1st Cavalry Division (Airmobile)

In March 1965 the tentative decision was made to convert the 11th Air Assault Division (Test) to a full-fledged member of the force structure. General Creighton W. Abrams, who was then the Vice Chief of Staff, said after the decision briefing, "Is it not fortuitous that we happen to have this organization in existence at this point in time?" Those who had been fighting for such an organization for over a decade could not help but sense the irony of this remark. It was decided that the new division would carry the colors of the 1st Cavalry Division which was then deployed in Korea. This decision was made for a variety of reasons, some of them emotional and some pragmatic.

On 1 July 1965 the 1st Cavalry Division (Airmobile) was officially activated pursuant to General Order 185, Headquarters Third U. S. Army, and was made up of the resources of the 11th Air Assault Division (Test) and the 2d Infantry Division. Despite a crippling loss of personnel by reassignment throughout the division, its personnel were able to retrain, re-equip, and deploy this major force to combat in ninety days. This effort is a major story in itself. Almost 50 percent of the original personnel were ineligible for overseas deployment. Replacement pilots had to be trained on new aircraft and new standing operating procedures, and the original structure itself received major modification. For example, the

Mohawk attack aircraft had been eliminated [2] (six Mohawks were retained for reconnaissance and surveillance). A full brigade of the division were to be qualified paratroopers, the Little John battalion had been deleted, and the aviation group had been drastically modified. A chart of the 1st Cavalry Division as deployed is shown on page 59.

The division staged out of Mobile, Alabama, and Jacksonville, Florida, on the USS *Boxer* and three Military Sea Transportation Service ships. Approximately 80,000 man hours were required to process all the aircraft aboard the four vessels. The USS *Boxer* proceeded via the Suez Canal while the three smaller vessels traveled the Pacific.

An advance party landed in the Republic of Vietnam on 25 August 1965 and arrived at An Khe shortly thereafter. (*Map 1*) The men immediately began clearing the "golf course," which was to become the world's largest helipad. The 1st Cavalry Division was about to write a new chapter to its proud history.

---

[2] The air assault division included 24 armed Mohawks in its table of organization and equipment. Nothing could have raised a brighter red flag in front of those proponents of complete Air Force control of all aspects of air support. An unbelievable amount of time and effort was devoted to the roles and missions aspect of the Mohawk during this period. General Johnson, Army Chief of Staff, remarked during the course of the air assault tests that he devoted more than 60 percent of his time in the joint arena to this one relatively small system. Indeed he felt that the whole future of the Army airmobile concept might be jeopardized by the Army's unwarranted concentration on hanging a few machine guns on a twin-turbine airplane. General Johnson described the Mohawk as the "barber pole of the air assault division." After being asked what he meant by this phrase, he replied that when one looks for a barber shop he finds it by looking for a barber pole; and one could find an air assault division in a theater by looking for the armed Mohawks.

CHAPTER IV

The First Airmobile Division and the Buildup, 1965

Buildup of U. S. Ground Forces

General Westmoreland, Commander Military Assistance Command, Vietnam, in his report on the war in Vietnam, stated:

> The year 1965 was one of momentous decisions and of commitment . . . In making my recommendations in the spring and early summer of 1965, as indeed in the case of later recommendations, I was mindful of the stated U. S. objective with respect to Vietnam: "To defeat aggression so that the people of South Vietnam will be free to shape their own destiny." It was my judgment that this end could not be achieved without the deployment of U. S. forces. With the concurrence of Ambassador Taylor, I so recommended.

The United States had already made a large commitment of airmobility assets to Vietnam in support of Army of the Republic of Vietnam forces. Now U. S. ground forces would test the airmobility concept for the first time in combat. Propinquity dictated that the first major combat unit of the U. S. Army to be deployed would be the airborne brigade stationed in Okinawa. On 5 May 1965 the 173d Airborne Brigade, comprised of two battalions of infantry and one of artillery, arrived from Okinawa to provide security for the major air base at Bien Hoa and the airfield at Vung Tau. It would not be long before this brigade would be committed to major offensive action.[1]

Brigadier General Ellis W. Williamson, the commander of the 173d, had taken immediate action to prepare his troops for the peculiar environment in Vietnam and especially highlighted their training in counter-ambush. By 10 May 1965, he noted:

> We have started our airmobile training. At first we are just practicing "getting in and getting out" techniques. As far as the individual is concerned, this is a critical operation. Get in quickly and get out

---

[1] A Marine Expeditionary Brigade had arrived in Da Nang on 8 March with the primary mission of securing the large air base in the northern province.

quickly and move as rapidly as possible from the landing zone in the right direction into the woodlines. This movement must be made behind a blast of fire from our own hand-carried weapons.

Two days later in a note to the troops, General Williamson said:

> The helicopter fire support gunships often frighten new troops. They whip over your heads rather quickly, firing on your flank or in front of you. Don't be fooled by the falling expended cartridges or belt links. Often inexperienced troops mistakenly think that these cartridges and links falling among them are fire from the helicopter. They won't hurt you. The real bullets are on the target. These helicopter pilots are some of the most efficient, professional men that we have ever observed. If we can identify the target for them they can and will hit it.

During the month of June the 173d was joined by the 1st Battalion of the Royal Australian Regiment. The Australians would be working directly with the Brigade for some months and would soon be joined by a field artillery battery from New Zealand.

On 28 June the 173d Airborne Brigade participated in the largest troop lift operation conducted in the Republic of South Vietnam up to that time. Over 144 Army aircraft, including 77 troop transport helicopters, lifted two battalions of the Vietnamese 2d Airborne Brigade and the 1st and 2d Battalions of the 503d Infantry deep into War Zone D. In all, nine battalions were involved: five infantry; one artillery; one support; a composite battalion of cavalry, armor, and engineers; and the Australian battalion. This was the first time any large-size force of friendly troops had operated in this area for more than a year. Many caches of weapons and rice were destroyed. Twenty-five Viet Cong were killed and fifty or more wounded. On D+2, the forces were extracted.

On 6 July the 173d returned to War Zone D and conducted one of its most successful operations since its arrival in Vietnam. In conjunction with the battalion of Australians and units of the 43d Army of the Republic of Vietnam Regiment, multiple air assaults were made just north of the Song Dong Nai River. Some 1,494 helicopter sorties were flown in support of this operation. Fifty-six Viet Cong were killed by actual body count. Twenty-eight prisoners of war were taken and one hundred tons of rice plus literally tons of documents were captured.

A Critique of an Air Assault

One of the most valuable tools of a commander is the immediate review of a combat operation with his subordinate commanders and staff. With the events fresh in mind, they discuss what went right and what went wrong–not for the historian, but rather to improve the next operation. In the earlier U. S. operations these critiques were particularly important, as is shown from the following excerpts of the meeting after the assault of the 173d into War Zone D.

In his critique of the operation, General Williamson said:

> In all candor I must admit that I did not expect to find as many enemy in that area as we did . . . We did a lot of things that we could not even have considered six weeks ago. As you recall when we first arrived in Vietnam we started off doing one thing at a time. On this operation, at the extraction time, we took 3,000 troops out of three different landing zones in three hours and ten minutes. We wouldn't have moved troops that fast or afford to bring our troops that close together at one time unless we had a lot going on at one time . . . As I looked at it from above, it was a sight to see. We were withdrawing from the center LZ while some friendly troops were still in the western LZ. We had a helicopter strike going in a circle around the center LZ. The machine gun and rocket firing helicopters kept making their circle smaller and smaller as we withdrew our landing zone security. Just to the west side we had another helicopter strike running north to south. We also had something else that was just a little hairy but it worked without any question, the artillery was firing high angle fire to screen the northern side of the landing zone. The personnel lift helicopters were coming from the east, going under the artillery fire, sitting down in the LZ to pick up troops and leaving by way of the southwest. In addition to that, we had an air strike going to the northeast. All of these activities were going on at the same time. We could not have done that a few weeks ago.

The 173d found that they had to go to unusual lengths to clear new landing zones for medical evacuation. The commanding officer of the 1st Battalion (Airborne), 503d Infantry, told how one of his company commanders tried to hack out an emergency landing zone in the jungle. "They used over 100 pounds of C–4 explosive that they had with them, and they had a couple of axes and hacked away with their machetes. The C–4 took the bark off and made a few splinters, and that was about it. These trees were almost totally resistant to the charges."

The S–3 of the 1st Battalion of the Royal Australian Regiment, noted one problem as he saw it in working with the helicopters. "I would like someone to make an assessment to what extent do we lose the initiative by the excessive use of helicopters. By the use of them, the enemy can determine where you are, and the strength you are in."

The Air Force liaison officer remarked:

We benefited from this operation in that we faced our close air support problems at last in practice, rather than in theory. The FAC's gained experience—just as in your profession—the classroom is a far cry from actually delivering weapons in close proximity to friendly troops . . . We have uncovered a number of in-house problems which we are busily engaged in correcting. On the strength of what we have learned, we can promise you far better air support in the future.

Lieutenant Colonel Lee E. Surut, Commanding Officer of the 3d Battalion, 319th Artillery, closed the critique on a wry note: "Gentlemen, we are rapidly approaching the time when the critiques will be as long as the exercises. Once again the artillery lent dignity to what would otherwise have been a vulgar brawl . . . The artillery fired 4,857 rounds of 105mm ammunition during the four-day exercise . . . The Artillery put a lot of bullets on the target and accounted for significant destruction and VC casualties."

Aviation Support

The early experience of the 173d Airborne Brigade pointed out some of the advantages of attached airmobility. During its first three months in Vietnam the brigade was supported by the 145th Aviation Battalion. In September, a helicopter company was attached and collocated. Aviation support by attached aviation was so much better than support on a mission basis that the brigade staff was highly laudatory of the aviation company and the advantages of attachment.

A study of this relationship was made by the Combat Developments Command during this period. The study concluded:

The cohesiveness and teamwork developed between the supported and supporting units is extremely important. The aviation company is attached to the brigade, lives with the brigade, and works with the brigade on a daily basis. They are a part of the brigade. This unit can move a battalion anywhere in the TAOR within two hours. It would take two or three times as long if you had to go to an aviation unit outside the brigade. This closeness and cohesiveness between the brigade and the aviation company has been achieved through constant practice and improvement of airmobile techniques. Teamwork is the key word.

Furthermore, effectiveness of the attached company increased appreciably. After attachment and collocation, the total sorties flown per month by the aviation company *increased* 24 percent, the number of combat sorties *increased* 165 percent, and the average tonnage carried per month *increased* 50 percent; while the average number of hours flown monthly per pilot *decreased* 24 percent and the average number of hours each aircraft was utilized *decreased* 23 percent. These data are based on a comparison of data developed over a six-month period, three months in support and three months attached.

The early operations of the 173d demonstrated the absolute necessity of "orchestrating" an air assault operation. As General Williamson had pointed out, an airmobile operation was no simple matter of moving troops from point "A" to point "B" if you really wanted to exploit the potential of the helicopter. It took training and time to integrate tactical air, helicopter gunships, field artillery, reconnaissance, and troop maneuver elements into a single swift operation.

Growing Pains

The early operations of the 173d would not have been possible had not the helicopter assets been in place and trained before their arrival. During this time we were "robbing Peter to pay Paul" throughout our worldwide inventory. In Vietnam as U. S. troop strength grew, we had to reduce our helicopter support to Army of the Republic of Vietnam forces to give priority to the U. S. Army ground units. It would not be until the arrival of the 1st Cavalry Division that any significant number of organic helicopters would arrive with their parent unit.

On 20 July 1965 the U. S. Army Support Command, Vietnam was redesignated U. S. Army, Vietnam, and General Norton was designated Deputy Commanding General. This change was indicative of the growing presence of U. S. ground troops and the necessity for better command and control procedures. On 28 July, President Lyndon B. Johnson announced that our forces in Vietnam would be raised from 75,000 to 125,000 and that additional forces would be sent as requested. On 29 July, the 1st Brigade of the 101st Airborne Division arrived in Vietnam following a brigade of the 1st Infantry Division which had arrived a few days earlier. The Marine strength continued to grow in the I Corps Tactical Zone as well as the theater logistical base to support the U. S. buildup.

The problems involved in this buildup can be described by a short review of the deployment of the 1st Brigade of the 101st Airborne Division. During the period 6 July to 29 July 1965 the Brigade moved from Fort Campbell, Kentucky to Vietnam. On 29 July through 21 August 1965 the Brigade manned a defensive perimeter in the Cam Ranh Bay area and began to establish a base camp. From 10 to 21 August the Brigade conducted operations southwest of Nha Trang and on 22 August, the Brigade moved north by sea and air with a mission to sweep clear the An Khe area of Binh Dinh Province to provide security for the arrival of the

1st Cavalry Division (Airmobile). During this deployment and subsequent operations, the Brigade had reconfigured to a new Table of Organization and Equipment and conducted major training in airmobile operations. They found that many of the items that they had brought in country such as camouflage nets were completely extraneous and many things they had not brought, such as additional water trucks, were absolutely essential.

When the Brigade was alerted at Fort Campbell, there was no positive assurance this would be a permanent move or a temporary duty move, and personal problems of families, quarters, etc., placed an undue administrative burden on the unit. The advance party moved on two old C–124 aircraft which, because of mechanical and administrative difficulties, required seven days to close in Nha Trang. The main body of over 3,600 troops was shipped aboard the USNS *General Leroy Eltinge* which had a normal troop capacity of 2,200 men. The 21-day voyage was a miserable period plagued by plumbing, lighting, and ventilation problems.

The movement of the 1st Brigade, 101st, to secure the An Khe base area for the soon-to-arrive 1st Cavalry Division was entitled Operation HIGHLAND and spanned the period 22 August until 2 October. One battalion conducted an airmobile assault in conjunction with a battalion-size ground attack to open the An Khe Pass and to clear and secure Route 19 from Qui Nhon to An Khe. To secure the division base area, the Brigade conducted eight airmobile assaults and many large ground operations. A special task force was organized to secure convoy movement along Route 19. This task force established strong points along the critical terrain bordering the route from Qui Nhon to the An Khe Pass. Tactical air cover was provided for all convoys. During this period, enemy losses totaled 692 killed in action as opposed to the 1st Brigade losses of 21.

Deployment of the Cavalry

Meanwhile, the newly designated 1st Cavalry Division was feverishly preparing for deployment. The movement of over 400 aircraft, nearly 16,000 personnel, over 1,600 vehicles, and training for combat in just eight weeks was a momentous task.

The early model Chinooks needed 1,334 modifications for safety of flight and the Hueys nearly 2,000. New armored seats were installed in the UH–1D's and the aircraft were moved to two different staging areas at Brookley Air Force Base, Mobile, Alabama, and Mayport Naval Base near Jacksonville, Florida. The USS

THE BOXER, LOADED WITH 1ST AIR CAVALRY AIRCRAFT, JUST PRIOR TO DEPARTURE FOR VIETNAM

The Boxer Leaves for Vietnam

Boxer and three Military Sea Transportation Service ships had been designated to move the division. All the Hueys were cocooned with "spraylat" and the other aircraft had separate preservative techniques applied. As mentioned earlier, a "crash effort" was required to process all the aircraft aboard the four ships. Since only the USS *Boxer* had adequate fire-fighting and servicing equipment aboard, Army equipment had to be acquired for shipment aboard other vessels to insure adequate fire protection and servicing prior to off-loading at destination.

One tends to forget the strategic mobility problems in moving such a force as the 1st Cavalry Division throughout the world and concentrates on the tactical mobility of the smaller units. This deployment was a non-tactical move and the Division did not land ready to fight. The following firsthand account by Colonel Stockton, who commanded the Air Cavalry Squadron, gives special insight into this problem:

About half way across the Pacific Ocean, I received a cable from the commander of the divisional task force to which my squadron was assigned on two of the three contingency plans to become effective on arrival in Vietnamese waters. Following instructions he had in turn received from high headquarters, Colonel Ray Lynch directed me to be prepared to fight my way ashore!

I remember to this day the stupefaction with which I received Colonel Lynch's message. Like the remainder of the division, my squadron had been loaded administratively. Our 90 helicopters were on board three freighters and a Navy aircraft carrier. Similarly, our 120 wheel vehicles, both combat and administrative, were spread among another half dozen Liberty ships which were God knows where in the Pacific. All I had on board the troop ship with me was the bulk of my officers and men, together with their individual weapons—perhaps a total of 600 people, or three quarters of the actual strength of the squadron. The rest were parcelled out on the other ships with our equipment, guarding it and preparing to offload it on arrival in Vietnam.

When in July I had been informed that we would be shipped to the combat theater of operations in the most convenient manner instead of the most tactically sound manner, I screamed loudly and vociferously, along with every other commander in the division, to anyone in authority whose ear I could catch. General Kinnard was as concerned about this as the rest of us, but we were all defeated by the machinery of inertia at those reaches of the Defense Department concerned with the military shipment of people and things.

Regardless of the fact that we had no tools with which to fight as cavalrymen, my orders from Colonel Lynch were explicit. Together with my valiant troop commanders, I worked out a scheme for making an assault landing in the Qui Nhon area with the assets we had on hand.

This done, I made a call on the ship's master, informing him of the instructions I had received and requesting that he break out the disembarkation nets we would have to use to get over the side so that we could practice with them. Here I was stupefied for the second time within a matter of less than 24 hours. Not only were there no assault landing nets on board, but the master had not even been informed of his sailing destination! He honestly thought that we might be going to Korea or possibly to the Philippines. In either event, he was sure that he would be tied up to a dock for unloading in the usual fashion. I finally persuaded this splendid seaman that we were in fact headed for battle-torn Vietnam. Neither he nor any of his officers had sailed in those waters for a dozen years. They were astounded at the prospect and assumed that their destination would be some location where adequate dock facilities existed for discharging their cargo. I was altogether unable to convince the master that we were in fact headed for Qui Nhon harbor where no unloading facilities of any kind were available.

Fortunately, as it turned out, our disembarkation was conducted peacefully and without interference by the simple procedure of using shallow draft vessels to lighter us ashore from our anchorage. Had it been otherwise, though—had there been trouble when we arrived—the 1st Cavalry Division could easily have been decimated before even a soldier of its main fighting component set foot on dry land. Even from the vantage point of more than two years of hindsight, I still shudder at the recollection.

The An Khe Hub

The 1st Cavalry Division had dispatched by Air Force transport a 1,000 man advance party under the command of Brigadier General John M. Wright, Jr., which arrived in An Khe on 27 August. As mentioned earlier, security for the advance party was provided by the 1st Brigade, 101st Airborne Division. September was spent in clearing away the semi-jungle area that was to become known as the "Golf Course."

When the ships with the main body arrived at Qui Nhon, the aircraft prepared for flight while on board the carriers. Only the Mohawks and two non-flyable Hueys were off-loaded by floating crane. The Chinooks quickly became the prime movers for troops and equipment between Qui Nhon and An Khe.

Although the Division did not completely reach An Khe until 3 October, it had assumed responsibility for its own security on 28 September, and elements of the division had already seen combat in support of the 1st Brigade of the 101st. In the first week of October, the 1st Cavalry was assigned a tactical area of responsibility which was essentially a private hunting domain in which the Division could operate by simply notifying any friendly forces that were in that area. This area soon grew to a zone approximately 150 by 150 miles. The 3d Brigade was given the populous coastal province of Binh Dinh, the 2d Brigade Kontum Province, and the 1st Brigade operated in the highland province of Pleiku which included the Chu Pong Mountains. The Cavalry Squadron was charged with a wide-sweeping reconnaissance mission throughout the entire area, but most particularly around the Special Forces camps at Pleime, Duc Co, Plei Djereng, Plei Murong, and Dak To that dotted the western highlands.

Major General Harry W. O. Kinnard recognized the vulnerability of the base camp at An Khe and consequently made it as small and compact as it could be so that it could be defended with a minimum number of forces. A 100 meter cleared perimeter completely encircled the camp with obstacles and surveillance devices. Artillery was positioned to develop quick fire concentrations on all likely enemy mortar sites and infantry patrols and ambushes were habitually employed out to enemy mortar range. One Air Cavalry troop was normally retained in the base camp area and aerial rocket artillery aircraft were kept on night strip alert to augment the fire of conventional artillery.

The establishment of the An Khe "hub of the wheel" for the

early 1st Cavalry operations was done for several reasons, not the least of which was the maintenance requirement.

It has often been said, with a certain amount of truth, that the Army refused to face up to the price that must be paid for airmobility. As a consequence, its organizations have usually been short of the necessary maintenance, supply, and security personnel. Part of this chronic shortage resulted from a long-standing battle between the "user" and the logisticians. Understandably since the early concept of an airmobile division, the tactical commander has wanted organic maintenance detachments down to the battalion level. This gives him the maximum responsiveness and a great deal of flexibility. From a logistician's viewpoint, such decentralization is a fragmentation of scarce skills and expensive special tools.

The tests of the 11th Air Assault Division included an extensive evaluation of what was known as the A–B–C maintenance concept. In this concept, the A-level was organic to the battalion and was designed to do all that repair which would return a helicopter to operable status within four hours. The B-level was organic to the division and could do all echelons of maintenance short of depot repair. The C-level provided that echelon of aircraft repair beyond the capability of the direct support levels.

While the 11th Air Assault tests conceded that the A–B–C system worked, the logisticians convinced the Department of the Army that it was impractical to field the 1st Cavalry Division in the time required using this maintenance concept. As a consequence, An Khe became a sophisticated fixed maintenance and supply depot which would inhibit tactical planning for some time to come.

The Ia Drang

By mid-October 1965, the North Vietnamese Army had begun its major operation in the Central Highlands. There is every reason to believe that it planned to cut South Vietnam in two at this time, for three North Vietnamese Army regiments had assembled in western Pleiku Province and adjacent Cambodia. On 19 October, the enemy opened his campaign with an attack on the Pleime Special Forces Camp twenty-five miles southwest of Pleiku. On 27 October, General Westmoreland directed General Kinnard to move his 1st Cavalry Division and seek out and destroy this enemy force consisting of the 32d, 33d, and 66th North Vietnamese Army Regiments. This became the month-long campaign known as The Battle of the Ia Drang Valley.

Initially the 1st Cavalry Division reinforced the South Vietnamese Army in relieving the Pleime Camp, and the North Viet-

JUMPING FROM A HUEY HELICOPTER

namese Army regiments broke contact and disappeared into the jungle. Little was known about the enemy's direction of movement except to speculate that they had gone west toward the Cambodian border. The 1st Brigade of the 1st Cavalry was given the mission of organizing a systematic search for the elusive enemy.

It was apparent that the Pleime Camp had been hit—and hit hard—by the enemy and it seemed inconceivable to the Air Cavalry Squadron that thousands of Viet Cong and North Vietnamese soldiers could completely disappear. On 1 November, Captain William P. Gillette, the Air Cavalry Squadron intelligence officer, spotted some unusual activity just eight kilometers west of the Pleime Camp and the Squadron was quick to capitalize on this information. Before the day was over, the Cavalry Squadron had committed most of its rifle and gunship platoons into the skirmish that developed. They killed 78 of the enemy and captured 57 prisoners, all of whom were regular North Vietnamese soldiers carrying identity cards issued in Hanoi. During the encounter the squadron lost five troopers killed and another seventeen wounded. It was evident that the Cavalry had uncovered a major enemy hospital area since case upon case of penicillin, quinine, morphine, and other valuable drugs lay piled four feet high over an area of more than a hundred square meters. Three complete sets of surgical instruments were mixed in with the medicine.

In this first skirmish of the Ia Drang Valley campaign, the Cavalry troopers were initially taken back at the almost suicidal short ranges at which they came to grips with the enemy. The bulk of the enemy attack force was within less than twenty meters of their perimeter before it was discovered and the enemy's "bear hug" tactics made supporting fires extremely difficult to place safely. Emergency medical evacuation landing zones had to be literally hacked out of the jungle with hand axes. Infantry reinforcements arrived at the site too late to take full advantage of this enemy contact.

The next day Brigadier General Richard T. Knowles, the Assistant Division Commander, met with Colonel Stockton to plot the strategy for further exploitation of this contact. The best estimate was that the major enemy force had moved along the Ia Drang Valley close to the base of the Chu Pong Mountains and the 1st Squadron, 9th Cavalry was given the mission of establishing an ambush in this area. A site was chosen called landing zone MARY, and the Cavalry Squadron fought its major battle in this area.

Landing zone MARY was unique in that it was the first time that the 1st Cavalry Division had mounted a successful night ambush and reinforced their attack with a night lift of an Infantry company. Also, they had developed their fire procedures to the point that armed helicopters were able to fire within fifty meters of the friendly troops during night operations.

By 4 November, the Cavalry Squadron had developed the battle to a point where it could be turned over to the infantry battalions. Over 150 enemy casualties had been accounted for with a loss of 4 U. S. soldiers killed and 25 wounded. The Ia Drang Valley Campaign was now fully underway.

On 9 November the 1st Brigade was relieved by the 3d, commanded by Colonel Thomas W. Brown. The 3d Brigade was given the mission of continuing the search south and southeast of Pleime. Colonel Brown continued his search in the densely wooded area south of the Ia Drang River at the base of the Chu Pong massif. He decided that the 1st Battalion, 7th Cavalry was in the best position to make the initial air assaults.

On 14 November the 1st Battalion, 7th Cavalry, commanded by Lieutenant Colonel Harold G. Moore, began the pivotal operation of the Ia Drang Campaign. He had chosen landing zone

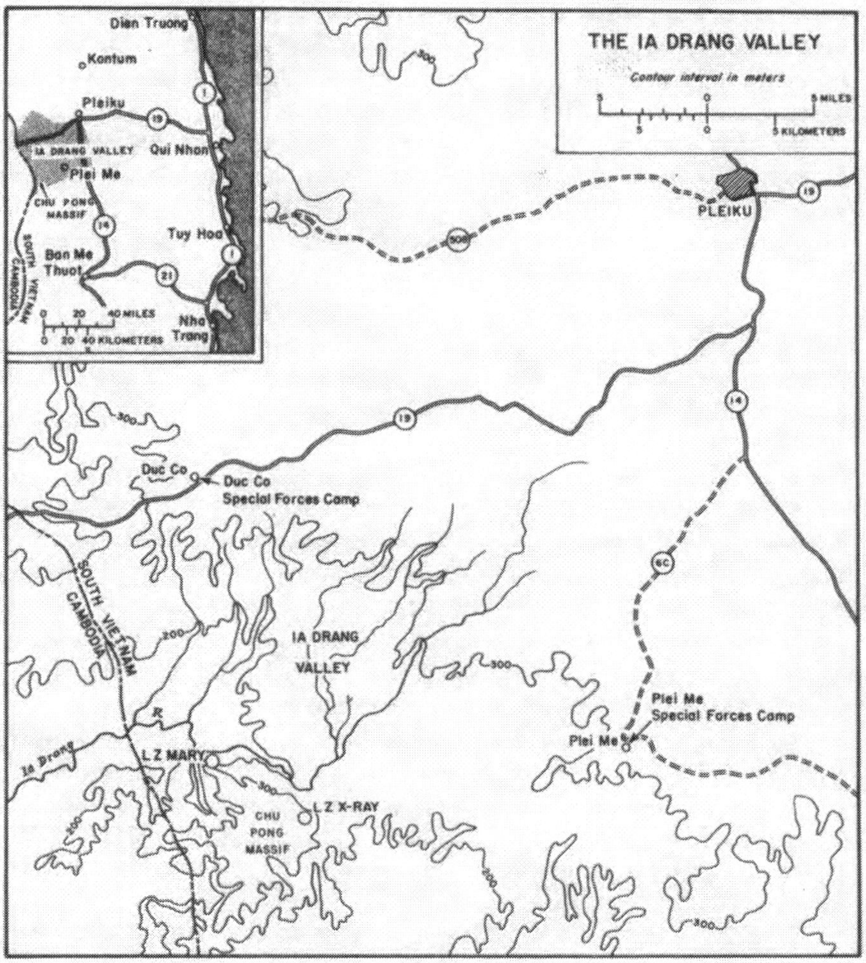

MAP 2

X-RAY (*Map 2*) out of the possible landing zones as the best potential position for the initial air assault. The cavalry section had confirmed that landing zone X-RAY could take eight to ten UH-1D's at one time. No signs of enemy activity were detected. Two artillery batteries were in position to support the landing.

Preparatory fire began at 1017 hours precisely where required and was timed with the lead elements of the assault company. The aerial artillery came on the heels of the tube artillery fire and worked over the area for 30 seconds expending half their load, then went into a orbit nearby to be on call. The lift battalion gunships took up the fire and were immediately ahead of the troop transport Hueys. The terrain was flat with scrub trees up to 100 feet high, thick elephant grass varying in height from one to five feet, and ant hills up to eight feet high. Along the western edge of the landing zone, the trees and grass were especially thick and extended through the jungles to the foothills of the Chu Pong Mountains. Company B made the initial assault.

The commander of Company B, Captain John D. Herren, secured the landing zone by having his 1st platoon dispatch its squads into different areas 50 to 100 meters off the landing zone to reconnoiter while he retained the balance of his company concealed near the center as an offensive striking force. The first prisoner, taken at 1120, stated that there were three battalions of the enemy on the mountain above the landing zone who were just waiting for an opportunity to attack. Company A, commanded by Captain Ramon A. Nadal, II, followed Company B into the landing zone unopposed and the perimeter expanded.

At 1330, B Company reported that it had been heavily attacked by at least two companies and its 2d platoon was in danger of being surrounded and cut off. Then 60-mm and 81-mm mortar fire began falling in the landing zone and on Company B. Company C arrived and its commander, Captain Robert H. Edwards, was ordered to take up blocking positions to the south and southwest of the landing zone to protect it from being overrun from that direction. The company commanders, the forward observers, the forward air controller, and the artillery liaison officer were all having difficulty getting coordinated as to the location of the forward elements of the company. There was no well-defined terrain feature to help identify positions and the air was heavy with smoke and dust. Company B seemed to be in the worst position since it had one platoon separated from the rest of the company in the jungle and could not precisely pinpoint its location to bring in fire support. By midafternoon the battalion knew it was in a major battle and fighting for its very existence. The enemy was coming from all sides.

As the lead elements of D Company landed, the helicopters took numerous hits, but none was shot down. One radio operator was killed before he could dismount from the helicopter and the door gunner and pilot were wounded. Colonel Moore stopped the other eight UH–1D's from landing by radio. Those who had landed from D Company immediately became engaged in the fire fight near A Company. The fighting became more intense. Colonel Moore decided to pull back A and B Companies under the cover of heavy supporting fire and smoke to the fringe of the landing zone and set up a tight defensive perimeter for the night. White phosphorus artillery was brought in and caused a temporary lull in the enemy firing that enabled some of the friendly forces to retrieve their dead and wounded and regroup. Both A and B Companies had numerous wounded and killed in action, while C Company had taken a few casualties but was in good shape. Company D had hardly been hit. During the afternoon Colonel Moore asked for assistance. The only company immediately available was Company B, 2d Battalion, 7th Cavalry, which landed in the landing zone by 1800. By 1900, the organization of the perimeter was completed. Units were tied in for the night and defensive artillery and mortar fires were registered.

A major problem that had developed throughout the day was the care and evacuation of the wounded. The battalion surgeon, medical supplies, and four aid station personnel had arrived at 1400 hours and, under heavy fire, they treated the wounded in the command post area. Colonel Moore had decided to restrict medical evacuation helicopters from coming in too frequently because most of the afternoon the landing zone was under fire. He had kept one small area big enough to land two helicopters open as his supply and evacuation link to the rear and intended to defend it at all costs. A system was arranged whereby every helicopter coming in with troops and supplies would have a specific load of wounded keyed to its departure.

Concerning his helicopter support, Colonel Moore stated:

> I have the highest admiration, praise and respect for the outstanding professionalism and courage of the UH–1D pilots and crews who ran a gantlet of enemy fire time after time to help us. They never refused to come in; they followed instructions beautifully; they were great. We in turn called them in when fire was the lightest and tried to have everything ready for each landing to keep them on the ground a minimum time. None were shot down and destroyed, although most of them took hits. Two aircraft were brought in which did not get out. One received enemy fire in the engine and had to land in an open area just off the northern portion of the LZ; the other clipped a few tree tops with the main rotor upon landing in the LZ and had to be left. Crews of both aircraft were immediately lifted out by other helicopters. Both downed helicopters were immediately secured by elements of Company D without orders per battalion SOP—they being the nearest troops. Both were slightly damaged only and were slung out two days later by CH–47 Chinooks. During the three day battle these were the only two downed helicopters.

By late afternoon it had become apparent that the battalion would need a night landing capability and Company A of the 229th Helicopter Battalion had anticipated the requirement. A pathfinder team arrived and cleared a fairly safe zone with engineer demolitions and set up the necessary lights for night landings. This remarkable feat was accomplished under enemy observation and fire. By darkness at 1915 hours, a resupply of ammunition, rations, water, and medical supplies had been brought in even though smoke hung like a horizontal curtain over the entire area.

Early that night the wounded had all been evacuated and the dead had been collected in the command post area. Mortar and artillery fires were registered close to the perimeter and the battalion prepared for night attacks. Company B still had one platoon that was cut off and surrounded, but it was reported holding its position with good morale. Later it was learned that this platoon began and ended the night with eight killed in action, twelve wounded in action, and only seven men not hurt. They had lost their platoon leader, platoon sergeant, and weapons squad leader. The platoon received several attacks during the night by an estimated 50-man enemy force. All were beaten off by small arms fire and artillery concentrations. A tactical air strike was made under flareship illumination. This was the only illumination used all night since it exposed the men in the surrounding area. However, there was a fairly bright moon from 2315 hours onwards. When daylight broke, numerous enemy dead were seen in the surrounding area.

At brigade headquarters, Colonel Brown continued to assess the significance of the day's activities. He was pleased that the 1st Battalion, 7th Cavalry, had been able to hold its own against heavy odds, and with moderate casualties, but was convinced that the fight was not yet over. He radioed General Kinnard for another battalion, and was informed that the 1st Battalion, 5th Cavalry, would begin arriving at brigade headquarters the following morning.

In the early morning hours savage close range fighting went on throughout the battalion perimeter. There was considerable hand-to-hand fighting. For example, the 1st Platoon leader of Company C was found killed with five dead enemy around him in and near his command post fox hole. Nearby, one trooper was found killed in action with his hands at the throat of a dead North Vietnamese Army soldier.

At approximately 0715 hours, the enemy attacked the sector of Company D, near where the mortars were emplaced. This put the perimeter under attack from two directions. Artillery, aerial rocket artillery, and Tactical Air were called in and delivered accurate and extremely effective fires. The aerial rocket artillery literally rained the perimeter with its rocket concentrations. Shortly, Colonel Moore radioed brigade headquarters informing them of the situation and, in view of the heavy losses, requesting an additional reinforcing company. The request was approved, but the company was not brought in at that time due to heavy fire in the landing zone. The enemy fire was so heavy that movement toward or within the landing zone resulted in more friendly casualties. At 0755 all platoon positions threw a colored smoke grenade on order to define visually for Tac Air, aerial rocket artillery, and the artillery air observer the periphery of the perimeter. All fire support was brought in extremely close and a few rounds of artillery fell inside of the perimeter along with two napalm bombs. However, only two men were wounded from friendly fires. The troopers stood their ground and many individual acts of heroism were recorded. By 1000, the enemy attack had been defeated.

At approximately 1330, all companies in the perimeter screened out for 300 meters and policed the battlefield. The area was littered with dead enemy, weapons and equipment and there was massive evidence from the bloody trails, bandages, etc., that many other enemy had been dragged from the area. Some of the enemy dead were found literally stacked behind ant hills. Two prisoners were taken and evacuated.

Late in the afternoon of 14 November, the brigade commander had moved the 2d Battalion, 5th Cavalry, into landing zone VICTOR. At approximately 0800 hours the following morning, it headed on foot for landing zone X-RAY. By noon, it had closed on the position. With this added strength, the Cavalry troopers were able to attack early in the afternoon and reached the surrounded platoon at 1510. The platoon still had some ammunition left and their morale was good. The wounded were evacuated, and all units were disposed and dug in for the night.

Although the enemy strongly probed their positions during the early hours of 16 November, the Cavalry troops held their positions. Flares were used continuously. At first light, orders were given for all men on the perimeter to spray the trees, the ant hills, and bushes forward of their positions for just over a minute to kill infiltrators and snipers. Immediately upon firing, a force of approximately 30 to 40 enemy exposed themselves forward of Company A of the 2d Battalion, 7th Cavalry, and began firing their weapons. Apparently the "mad minute" of fire had prematurely triggered a planned enemy attack. The tactic of spraying had accounted for six enemy dead. Several dropped out of the trees in the surrounding areas. The men liked this particular tactic not only for the effect it had on the enemy but for the relief of tension that had been built up throughout the night. By 0930 hours, the first elements of the 2d Battalion, 7th Cavalry, began arriving at landing zone X-Ray and they closed by noon. Colonel Moore ordered one final sweep of the area before he turned over control of landing zone X-Ray to the new troopers. His troops were extracted by UH–1D's early in the afternoon without enemy opposition. Discipline was excellent, and the helicopters were only on the ground a few seconds because the aircraft loads had been pre-spotted. By 1830 hours all elements had closed into Camp Holloway, in the vicinity of Pleiku.

Colonel Moore, who was to receive his eagles in just seven days and thus be promoted away from his command, had every reason to be proud of the performance of his battalion. Some 634 enemy had been killed by actual body count and six had been captured. Seventy-nine troopers had been killed and 121 wounded. None were missing. The enemy had lost many small arms, heavy machine guns, and other equipment, and the individual U. S. soldier had acquitted himself well in the hardest kind of hand-to-hand battle.

In his after action report, Colonel Moore noted that aerial rocket artillery had been extremely effective. His commanders had confidence in bringing such fires extremely close to their own positions. He also had noted that tube artillery, aerial rocket artillery, and tactical air can be used at the same time without seriously downgrading the effectiveness of the fire or endangering the aircraft. The aerial rocket artillery and Tactical Air flew perpendicular to the artillery gun-target line in those cases when they were making a simultaneous attack on the same target areas. This technique was possible by close teamwork between the forward air controller and the artillery liaison officer.

As mentioned earlier, casualties had been a critical problem and attempts to help the wounded had caused additional casualties in attempting to get them out. Colonel Moore said, "I lost many leaders killed and wounded while recovering casualties. Wounded must be pulled back to some type of covered position and then treated. Troops must not get so concerned with casualties that they forget the enemy and their mission. Attempting to carry a man out requires up to four men as bearers which can hurt a unit at a critical time."

The 1st Cavalry troopers had found that the enemy was well trained, aggressive, and was equipped with a preponderance of automatic weapons with plenty of ammunition. He was an expert at camouflage and sought to close in strength to render U. S. fire support less effective and to force us to fight on his terms. He fought to the death. When wounded, he continued fighting and had to be approached with extreme care. Many friendly troops were shot by wounded North Vietnamese Army soldiers. He also appeared fanatic in his extreme efforts to recover bodies of his dead and wounded and their equipment.

In the after action report, high praise was given to the pathfinders and their control of the "hot" landing zone. Until they had arrived, every incoming aircraft had to be guided in and out by the battalion commander on the battalion command net. The pathfinders took over this responsibility and made possible the invaluable night supply and evacuation missions. The fact that pathfinders were available and trained as part of the 1st Cavalry Division's organization was testimony to the soundness of its early doctrine.[2]

The action of the 1st Battalion, 7th Cavalry, was one of a series of major actions fought by the 1st Cavalry Division in this, its first major campaign in Vietnam. The 3d Brigade continued its systematic search and destroy pattern and defeated each of the three enemy regiments in turn. All together, the 1st Cavalry Division and Army of the Republic of Vietnam troops killed an estimated 1800 North Vietnamese troops.

The battle lasted 35 days, and on 26 November 1965, the 1st Cavalry Division had completed its mission of pursuit and destruction. The statistics of the aviation units involved, to a large extent, tell the story of how the mission was accomplished.

During the 35 days of the campaign, the aircraft delivered 5,048 tons of cargo from the wholesale terminals to the hands of the troops in the field. In addition, they transported 8,216 tons into Pleiku from various depots (primarily Qui Nhon and Nha Trang). Whole infantry battalions and artillery batteries were

---

[2] The U. S. Army almost lost its "pathfinder" role completely after World War II. In the reorganization of the Service roles and missions, it was determined that the pathfinders for the joint airborne assaults should be Air Force personnel since the Air Force insisted its Combat Control Team should be inserted first and control all subsequent drops. However, in 1955, when I had the Airborne–Army Aviation Department at Fort Benning it became obvious that the Army should have its own pathfinder capability for the terminal control of the Army's organic helicopters. I received permission to reactivate Army training in these skills and Lieutenant Colonel John E. Stannard wrote the manual. Fundamental to this decision was a determination of where Army pathfinders would be assigned. We tried to incorporate the pathfinder skills into each infantry battalion, but this solution was disapproved by Continental Army Command. The pathfinder almost became a man-without-a-country until the formation of the 11th Air Assault Division, where they were made part of the Aviation Group. As a result, only the 1st Cavalry had organic, trained pathfinders when required. Our other units in Vietnam would find many occasions when they needed them—desperately.

moved by air, and approximately 2,700 refugees were moved to safety. In all this flying, 59 aircraft were hit by enemy fire—three while on the ground—and only four were shot down; of these four, three were recovered.

General Westmoreland stated, "The ability of the Americans to meet and defeat the best troops the enemy could put on the field of battle was once more demonstrated beyond any possible doubt, as was the validity of the Army's airmobile concept."

Overview of 1965

It has been well documented that by early 1965 the enemy had reinforced his units in Vietnam to the point of being able to move almost at will against major population areas. In fact, it looked as if the North Vietnamese Army were about to cut the country in two, right across the middle. It is also well known that the commitment of U. S. Forces stemmed the tide and turned it in the other direction. The major battles that took place were in the highlands where the 1st Cavalry answered the challenge of a powerful North Vietnamese Army division that was attempting to overrun the Special Forces camp at Pleime. This operation was in essence the watershed line; from that time on, the joint U. S. and Army of the Republic of Vietnam operations throughout the country regained the initiative and began to drive back the major enemy forces in an all-out counteroffensive.

In 1965 there were just not enough airmobility assets to go around. Partly by design and partly by default, most of the airmobile assets ended up in support of U. S. forces, to the detriment of overall Army of the Republic of Vietnam operations. The Army of the Republic of Vietnam units in general reverted to a mission of near-static ground operations close to the populated areas. The U. S. forces choppered into the jungles to fight the main force North Vietnamese Army units. During this phase, the Army of the Republic of Vietnam forces actually had less helicopters available to them than in the period before the buildup.

The first operations of the 1st Cavalry Division brought out many points which confirmed the basic conclusions of those who had worked with the airmobile concept over the years. Airmobility put a new dimension into ground warfare, but it did not change the nature of warfare itself and it certainly did not negate those basic rules of survival so important to the individual soldier.

The airmobile trooper, like the paratrooper before him, must be basically a professional infantryman, artilleryman, and so forth.

The advent of the helicopter does not permit him to neglect his individual weapon or other battlefield disciplines which have been essential to survival since recorded history. Airmobility, if anything, is particularly unforgiving of carelessness or sloppiness. The after action reports of airmobile units tend to repeat lessons learned in Korea and World War II on such things as ambushes, mines, and booby traps, rather than highlight new helicopter techniques. Indeed, some of the operations fought by the 1st Cavalry hardly make mention of their airmobility.

These records indicate in a way that these units took airmobility for granted and only mentioned aircraft technology when there was a need for improvement—just as jeeps and ¾ tons were seldom mentioned in dispatches of earlier battles. This is an indication that the helicopter was doing its job so well and so routinely that it was not considered worthy of mention.

This was not unique to the 1st Cavalry operations alone, for in writing this study, I was struck by the similarity of other units who used attached airmobile assets in a matter-of-fact attitude. To a large degree, these units considered themselves as much airmobile as the 1st Cavalry Division although they were not officially designated as such. This is a great credit to those separate airmobile companies and battalions who supported these units. Obviously they became so identified with the 173d, the 101st, and so forth, that they functioned with almost the same continuity as those helicopter units which were organic to the 1st Cavalry Division. This takes away nothing from the individual helicopter crew or the unit that they supported; but, there is a wealth of examples which demonstrated that airmobile expertise which had been carefully developed over a period of months suddenly was downgraded by the shifting of attachments. Every commander instinctively knew that he could do certain things with "his" Hueys that he couldn't quite do with "somebody else's."

By the end of 1965, it had become apparent that the "business-as-usual" approach to the aviation training base and helicopter production lines soon would create a major deficit in the Army's inventory of pilots and aircraft. Belated recognition of this fact produced an almost unmanageable surge in the pilot training program and a strain on every helicopter manufacturer's capacity, especially Bell. It would be almost two years before the aviation assets approached the Vietnam aviation requirements. Many Army aviators would find themselves faced with repetitive tours in Vietnam and many operations in Vietnam would be structured around the limitation of available helicopters rather than the more basic consideration of the enemy threat. Indeed, the management of aviation assets would soon become a major preoccupation of every senior commander.

CHAPTER V

Airmobility Comes of Age, 1966

Airmobility in the Delta

The 173d Airborne Brigade (Separate) launched the new year on 1 January 1966 with a smoothly executed move into the Mekong Delta. This was the first time an American ground unit had operated in the notorious "Plain of Reeds."

The Brigade had moved from Hau Nghia Province into the Delta by land and air. The air elements, consisting of the 1st Battalion (Airborne), 503d Infantry; the 2d Battalion (Airborne), 503d Infantry; the 1st Battalion, Royal Australian Regiment; and "C" Battery, 3d Battalion, 319th Artillery, came into Bao Trai airstrip.[1] By 1425 the 1st Battalion of the 503d had the honor of being the first American force to make an air assault west of the Oriental River. This assault, which was preceded by an effective landing zone preparation by Tactical Air, artillery, and armed helicopters, only experienced light opposition which was quickly brushed aside. The Australian battalion established themselves by air assault on the east side of the Oriental River which effectively cut this enemy supply route.

As previously scheduled, the 2d Battalion of the 503d remained at the brigade forward base until the following morning when they conducted a heliborne assault into landing zone WINE. This landing zone was approximately five kilometers south of the Australian position and also on the east side of the river. Here they met very strong enemy resistance and the battalion fought a bitter and fiercely-conducted battle throughout the day against a dug-in, well-concealed, battalion-size Viet Cong force. Intense artillery fire,

---

[1] The 2d Battalion, 503d Parachute Infantry Regiment was the first U.S. unit to jump in combat during World War II (Operation TORCH in Africa). As executive officer, I helped organize the 2d Battalion and later activated, trained and led the 3d Battalion of the 503d which made the first combat jump in the Pacific. As a coincidental footnote to history, when we made that jump into the Markham Valley of New Guinea, our total artillery support was a battery of "25– pounders" of the Royal Australian Artillery.

helicopter gunship fire, and Tactical Air pounded the Viet Cong positions continuously. Late in the afternoon a strong coordinated attack behind a wall of artillery fire drove the Viet Cong from his positions. He left 111 dead behind along with considerable equipment.

The Brigade attempted to employ the armored elements, Company D, 16th Armor and the Prince of Wales Light Horse (Australian armored personnel carriers), with the infantry battalions but found that the carriers had extreme difficulty in traversing the marshy rice paddies. They were returned to the Brigade base area and later were successfully employed on security missions. On D+2, the Australian battalion conducted search patrols south of their position and eventually linked up with the 2d Battalion of the 503d on the east bank of the Oriental River. Operations proceeded through D+4 with only light contact being made and the enemy obviously making every effort to remain hidden.

On D+5, the 173d Brigade demonstrated its growing sophistication in airmobile techniques. On this day, three Eagle Flights, one consisting of 144 men, the others 72 each, orbited the target area just beyond visual range. When the commander was ready to move the force in, he announced the command of execution "Skyhook," which was the signal for all supporting actions to include firing to begin a countdown. From this prearranged signal with minimum radio communications, all supporting actions were time-sequenced to place the landing force on the landing zone immediately after the coordinated delivery of supporting fires and a last minute reconnaissance. After the first flight landed, the second was in orbit position until it was apparent that no reinforcement was needed in the first landing zone. The "Skyhook" procedure was then reinitiated and the second flight landed on another nearby landing zone. Following the same procedure, a flight was put in on yet a third landing zone. By this maneuver, three stripped down companies had covered an area with reconnaissance by fire that normally would have taken a battalion or more, using previous formations.

Operation MARAUDER was terminated on D+7, 8 January 1966, after decimating the 267th Viet Cong Battalion and the headquarters of the Viet Cong 506th Battalion. The 173d Airborne Brigade, as the first American unit to operate in the Mekong Delta, demonstrated its ability to swiftly co-ordinate the tactical air, helicopter gunships, artillery, and troop maneuvers. They had ranged over a very large area but always had their artillery in position by Chinook movement. Armor had been used successfully when it was given proper terrain. Many procedures had now become standard to the point where complex operations, that would have required detailed rehearsal and briefings, could be done as a matter of routine.

Immediately following this operation, the 173d Airborne began Operation CRIMP in a drive through the Ho Bo Woods region in Binh Duong Province in an attempt to destroy the politico-military headquarters of the Viet Cong Military Region 4. After six days of tedious fighting through bunkers and tunnels, the mission was accomplished and the headquarters was found and destroyed. An enormous quantity of enemy documents and weapons were captured. The brigade commander, General Williamson, made special mention after this action of the role played by his support battalion. "The Support Battalion makes the 173d a truly separate brigade. Its performance has been exceptionally fine and represents a major contribution to our combat power by standing behind the infantry and those who support the infantry with supplies, transportation, maintenance and medical support." The Brigade was supported throughout these operations by the 145th Aviation Battalion (Reinforced). This battalion, seldom mentioned by name, had become almost completely identified with the 173d Brigade total force.

Airmobile Logistics

Even the most enthusiastic protagonist of the airmobility concept would readily admit that the introduction of hundreds of complex helicopters to a combat environment brought, along with its advantages, many unavoidable headaches, not the least of which was the maintenance requirement. In 1966 the helicopter was still in its adolescence from a technical standpoint. The yet unreached goal of one hour of maintenance per hour of flight was far in the future. In reality, even the simplest machine required approximately ten hours of maintenance per hour of flight. The fact that we were willing to pay that price was mute evidence of the intrinsic worth of airmobility.

The story of airmobility must include the fantastic individual efforts of crew chiefs and mechanics who worked practically every night, all night, to enable the helicopters to fight the following day. With the geometric increase in aircraft during the U. S. Forces buildup, it is important to take a brief look at the aircraft maintenance structure.

A milestone in aviation maintenance occurred with the formal approval of the organization of the 34th General Support Group by U. S. Army, Pacific General Order Number 6, dated 17 January 1966. This approval authorized a strength of 145 personnel for the Headquarters and Headquarters Company which, at that time, actually had a strength of only two officers and five enlisted men. This formal approval was the culmination of many earlier decisions on the best alternative solution to the growing aviation maintenance problem in Vietnam.

In July 1965 Army aircraft maintenance in Vietnam was provided by three direct support companies and one general support company. Aviation supply was managed by a special Aviation Supply Point in Saigon. All of these units were part of the 12th Aviation Group (Provisional) which in turn reported to the U. S. Army Support Command. At this time in 1965, about 660 Army aircraft were in Vietnam; but, plans were already firm to expand this aircraft population to over 2,000 by early 1966. General Norton, then Commanding General of U. S. Army Support Command, formed an ad hoc committee to study possible alternatives for dealing with this growing maintenance requirement. General Norton's objectives were to provide one-stop maintenance and supply support to Army aircraft (including airframe, engine, avionics and armament); and, to provide an organization that had the ability to grow with the requirement.

At that time, three organizational alternatives were possible. First, the aircraft maintenance and supply units could be integrated into the 1st Aviation Brigade structure. Second, these units could be integrated into the 1st Logistical Command structure. Third, a special separate command organization could be formed to control all non-divisional aircraft maintenance and supply units. This group could be assigned to either the 1st Aviation Brigade, the 1st Logistical Command, or directly under U. S. Army Vietnam Headquarters as a separate major command.

Major Rudolph D. Descoteau and Major Charles L. Smith, as members of the ad hoc committee, developed matrices which considered all alternatives including span of control, flexibility, and responsiveness. This committee also solicited opinion of their potential customers to include the 1st Cavalry Division. On 27 August 1965 the ad hoc committee submitted their new plan of the organization of aircraft maintenance and supply. This plan marked the conception of the 34th General Support Group though it was yet unnamed. A separate command organization to provide the aircraft maintenance and supply support seemed to offer the best solution. With this organizational structure the requirements to provide one-stop maintenance and supply support could be more easily satisfied. It was realized that certain duplication in their requisitioning systems would occur in common items, but the resultant responsive support warranted this approach. The next problem was to determine where the group should be placed in the overall command structure. Assignment to the Aviation Brigade would place it under the control of the operator for the highest aircraft density. However, this would reduce the probability of equitable support to divisional units and non-divisional–non-Aviation Brigade units such as signal and engineers. The assignment of the group to the 1st Logistical Command had the major advantage of concentrating logistical support under a single commander who would answer to the U. S. Army, Vietnam, G–4. The last alternative, having the group respond directly to U. S. Army, Vietnam, with no intervening layer, seemed in line with the critical nature of aviation assets.

The above alternatives were presented to General Norton and his staff by the ad hoc committee in September 1965 with a recommendation to adopt the separate group structure and have it report directly to U. S. Army, Vietnam. General Norton accepted the recommendation and directed implementation of the plan.

Recommendations for a management structure to provide aircraft support from a single-point, in-theater, aircraft repair parts inventory control center evolved into the Aviation Material Management Center concept. The Aviation Material Management Center Tables of distributions and allowances and the Combat Development Command Transportation Agency's proposed Table of Organization and Equipment for the General Support Group were forwarded for approval. U. S. Army Vietnam directed formation of a provisional group headquarters in November 1965. Lieutenant Colonel Ellis became the Group's commander and, using the assets of the old U. S. Army Support Command Aviation Detachment and the Aviation Supply Point, formed a skeleton staff. Personnel and equipment resource requirements were levied on the 14th and 765th Transportation Battalions to provide a minimal functional base.

The formal approval authorized the proper staffing of the headquarters but gave no relief in the critical day-to-day management requirements to expand the old Army Supply Point and establish a separate inventory control center to support the fast-growing aircraft fleet. The Supply Division of Group Headquarters performed this function until more help arrived. The 241st Transportation Company (Depot) arrived in February 1966, giving the Aviation Material Management Center the capability of operating two depots. In April the 58th Transportation Battalion arrived and assumed operational control of Aviation Material Management Center.

The Army's "Aircraft Carrier"

On 12 April 1966 the most unusual Army aviation maintenance battalion of the 34th General Support Group steamed into Cam Ranh Bay. The unit was the 1st Transportation Corps Battalion (Depot) (Seaborne), the only Floating Aircraft Maintenance Facility in the Army. The idea for this floating facility originated during military operations in the Pacific Theater during World War II, when combat areas switched rapidly from island to island and sudden changes in the combat zones made ground aircraft maintenance facilities almost useless.

As early as 1962 the floating aircraft maintenance facility concept was being developed for use in the Vietnam combat zone, but it was not until 1965 that the Navy seaplane tender USS *Albermarle* was actually selected for conversion to this facility. On March 27, 1965 it was rechristened the USNS *Corpus Christi Bay*. An energetic Army aviator, Colonel John Sullivan, scurried from the Pentagon to the shipyard to the Aviation Materiel Command to consolidate the many facets of this unusual undertaking. The Army, which had been accused by the Air Force of beginning another "Air Corps" now was getting strange looks from the Navy with its attempts to get its own "aircraft carrier."

When the red tape had been cut (lengthwise), the ship was modified to carry approximately 370 Army maintenance personnel and supporting technicians and 130 civilian maritime crewmen to operate the ship. Thirty-seven different production and support services were established aboard the ship enabling the facility to perform all maintenance functions of a depot level repair facility, including overhauling and rebuilding aircraft components. One of the most remarkable innovations was a technical data library on board which contained a complete file of 180,000 engineer drawings on film of aircraft systems, components, and special tools. In its library they had 785,000 images that could be broadcast throughout strategic areas on a closed circuit TV. In a sense, the ship represented an extension of the large aircraft maintenance facility at Corpus Christi, Texas, directly to Military Assistance Command, Vietnam.

When the *Corpus Christi Bay* first arrived at Cam Ranh Bay, extensive security precautions were taken to protect the ship from enemy action. The most serious threat was envisioned to be sabotage and guards periodically threw concussion grenades over the sides to discourage enemy underwater swimmers. A scuba team periodically checked the hull. On 21 September 1966 the *Corpus Christi Bay* moved out of Cam Ranh Bay and sailed to the harbor at Qui Nhon to be near the 1st Cavalry, the unit it primarily supported.

While the Floating Aircraft Maintenance Facility was a useful and unique addition to the U. S. Army's helicopter support capability, the more important, if less dramatic, support was performed in the open rice paddies and jungle clearings. There the maintenance personnel lived in constant danger, with practically none of the amenities, and performed daily, casual miracles on the complex aircraft. Never sure when the necessary bright lights would become an aiming point, they used the knowledge gleaned from Fort Rucker and Fort Eustis classrooms under the most primitive conditions. As one supervisor gruffly understated, "They done good!"

The 1st Cavalry Division in Binh Dinh

During the first half of January 1966 the 1st Brigade of the 1st Cavalry Division conducted Operation MATADOR to find and destroy the enemy in Pleiku and Kontum Provinces. During this operation, the 1st Cavalry saw the enemy flee across the border into Cambodia, confirming that the enemy had well-developed sanctuaries and base camps inside that country.

After Operation MATADOR, the 1st Cavalry Division shifted its weight toward Binh Dinh Province. Some of its forces had been committed into this area soon after its arrival in Vietnam in the summer of 1965, but the major effort in the Ia Drang Valley occupied most of the 1st Cavalry's attention throughout 1965.

The heavily populated rice plains in this area had a population of nearly half a million people of which at least 200,000 or more were still under the domination of the Viet Cong infrastructure. The South Vietnamese Government was attempting to extend its control north from Qui Nhon, along National Highway One, through the rich plains area up to Tam Quan in northeastern Binh Dinh. In this effort, the 22d Army of the Republic of Vietnam Division was being assisted by the Capitol Republic of Korea Infantry Division based near Qui Nhon. Only the southern part of the area was truly under government control.

Beyond the northern border of the Phu My District, and as far as the edge of the abrupt mountain range which walled off the plains on the north and west, there were only isolated islands of refugees. The rest of the rice bowl belonged to the enemy and was presided over by the 3d North Vietnam Army Division. Its three regiments—the 2d, the 18th, and the 22d—operated from mobile base camps hidden in the mountains. From there they sent small forces throughout the lowlands to terrorize the farmers, manipulate the cycle of rice growing and harvesting, and generally controlled the lives of the people of Binh Dinh.

The 1st Cavalry's initial major operation in this area was called MASHER in its first phase, and WHITE WING in its second, third, and fourth phases. The fighting covered a full circle around Bong Son. The 1st Cavalry Division, in close coordination with the 22d Army of the Republic of Vietnam Division, began with air assaults into the Cay Giep Mountains, then moved to the Bong Son Plains, the An Lao Valley, the Kim Son Valley, and finally back to the Cay Giep Mountains. As a result of MASHER-WHITE WING, the airmobile division and the Army of the Republic of Vietnam infantry forced the North Vietnamese Army regulars out of the area and temporarily broke their hold on the population. As it turned out, the 1st Cavalry would find itself preoccupied in this area on and off for a long time to come.

In the after action report of the 3d Brigade when it concluded Operation MASHER-WHITE WING on 17 February, they were able to report that 893 enemy had been killed by actual body count. A large quantity of equipment and small arms had been captured along with 24,000 rounds of ammunition. Friendly losses were 82 killed in action and 318 wounded.

The Brigade had been supported throughout this operation by the 133d Assault Support Helicopter Company with 16 Chinooks. The CH–47 Chinook had proved essential in moving artillery and resupplying the Brigade with ammunition and supplies. Night resupply was often required. On 28 January seven Chinooks made an emergency resupply mission during weather conditions consisting of extremely low ceilings and poor visibility, and six of the seven committed helicopters were hit by enemy ground fire. The company commander, Major Taylor D. Johnson, was killed while attempting to recover a downed OH–13 scout helicopter. Despite the weather and the enemy fire, the 16 Chinooks assigned to this company during the period 1 January through 31 January flew 526 hours transferring 3,212 passengers and over 1600 tons of cargo.

Lieutenant Colonel Max A. Clark, the commanding officer of the parent organization for this company, the 228th Assault Support Helicopter Battalion, made special note of the difficulty in supporting Chinook operations so far away from the An Khe base with the current shortage of pilots and maintenance personnel. It took a major effort of his entire battalion to maintain an availability rate of 58 percent.

The Role of the Chinook

The story of airmobility is essentially one of men and machines. If the Huey helicopter became the cornerstone of airmobility, then the Chinook must be considered one of the principal building blocks.

Late in 1956 the Department of the Army announced plans to replace the H-37 helicopter, which was powered by piston-driven engines, with a new, turbine-powered aircraft. A design competition was held and, in September 1958, a joint Army-Air Force source selection board recommended that the Army procure the Boeing Vertol medium transport helicopter. However, the necessary funds to proceed with full-scale development were not available and the Army vacillated in its design requirements. There were those in the Army who felt that this new helicopter should be a light tactical transport aimed at the mission of the old H-21's and H-34's and, consequently, sized for approximately fifteen troops. Another faction believed that the new transport should be much larger to serve as an artillery prime mover and have minimum interior dimensions compatible with the Pershing Missile system. This "sizing" problem was a critical decision.

The first Vertol prototype, called the YHC-1A, was tested by the Army to derive engineering and operational data. Three aircraft were built with a maximum troop capacity of twenty. This model eventually became Vertol's commercial 107 and the Marine Sea Knight. However, the YHC-1A was considered by most of the Army users to be too heavy for the assault role and too light for the transport role. The decision was made to procure a heavier transport helicopter and at the same time upgrade the Huey as a tactical troop transport. This decision was to determine the pattern of airmobile operations for the next decade. As a consequence, the Army concept of air assault operations differed from the Marines because, among many reasons, the very nature of the equipment demanded different methods of employment.

The "sizing" of the Chinook was directly related to the growth of the Huey and the Army's tacticians' insistance that initial air assaults be built around the squad. There was a critical stage in the Huey program when the technicians insisted that we should not go beyond the UH-1B model with Bell; that there should be a new tactical transport "between" the Huey and medium transport helicopter. Major General von Kann and I fought a rear-guard action in a Pentagon battle to keep the Huey program viable. When it was decided to go to the UH-1D (after an awkward pause on the original "C" design), the proper Chinook size became apparent. By resolutely pushing for the Huey and the Chinook, the Army accelerated its airmobility program by years.

The Army finally settled on the larger Chinook as its standard medium transport helicopter and as of February 1966, 161 aircraft had been delivered to the Army. The 1st Cavalry Division had brought their organic Chinook battalion with them when they arrived in 1965 and a separate aviation medium helicopter company, the 147th, had arrived in Vietnam on 29 November 1965. This latter company was initially placed in direct support of the 1st U. S. Infantry Division.

The most spectacular mission in Vietnam for the Chinook was the placing of artillery batteries in perilous mountain positions that were inaccessible by any other means, and then keeping them resupplied with large quantities of ammunition. The 1st Cavalry Division found that its Chinooks were limited to 7,000 pounds pay load when operating in the mountains, but could carry an additional 1,000 pounds when operating near the coast. The early Chinook design was limited by its rotor system which did not permit full use of the installed power, and the users were anxious for an improved version which would upgrade this system.

As with any new piece of equipment, the Chinook presented a major problem of "customer education." Commanders, pilots and crew chiefs had to be constantly alert that eager soldiers did not overload the temptingly large cargo compartment. I feel quite confident that Hannibal had the same problem with his elephants. It would be some time before the using troops would be experts at sling loads and educated in such minor details as removing the gunner's sight from the artillery pieces. The Chinook soon proved to be such an invaluable aircraft for artillery movement and heavy logistics that it was seldom used as an assault troop carrier. The early decision to move to this size helicopter proved to be indisputably sound.

Operation CRAZY HORSE

The origins of many of the major operations in Vietnam can be traced to some minor enemy contact which was quickly exploited by airmobile forces. Often this was the only way the elusive enemy

TROOPS BOARDING CH-47 CHINOOK HELICOPTERS

could be forced to fight. Operation CRAZY HORSE is a good example of the aggressiveness and determination of our forces in their search for the enemy. And it is appropriate that we examine the actions of the squad and platoon to understand how it came about.

The 1st Cavalry Division was finishing Operation DAVY CROCKETT on 15 May 1966 when a Civilian Irregular Defense Group (CIDG) patrol from the Vinh Thanh CIDG Camp, working the mountain valley immediately to the east, ambushed an enemy force and captured a mortar sight, 120-mm firing tables and a gunner's quadrant, plus some sketches of the CIDG camp and the hamlets in the valley. One company of the 1st Cavalry air assaulted into the hills east of the CIDG camp at 1000 hours on the 16th, to search out the area.

At 1100 hours on the 16th, Company B, 2d Battalion (Airborne), 8th Cavalry, commanded by Captain John D. Coleman, made a combat assault into what then was a one-ship landing zone named HEREFORD, a small patch of elephant grass about halfway up the side of the largest mountain east of the CIDG camp. Besides the usual forward observation party from the supporting artillery (an airmobile battery was landed adjacent to the camp), a Special Forces intelligence sergeant, his CIDG counterpart, and an interpreter moved with the company.

CH–47 Chinook Delivering 105-mm Howitzer (Towed) with Ammunition Pallet

After a hard climb to the ridgeline of the mountain, marked only by one pungi stake and two heat exhaustion casualties, the company began moving eastward along the razorback. At approximately 1400 hours, after having just crossed a slight rise in terrain, the lead platoon (the Third) spotted what appeared to be a single Viet Cong and opened fire. The fire was immediately returned in volume from prepared positions to the east. The platoon leader radioed Captain Coleman that he had encountered stiff opposition and was ordering a squad to begin a maneuver around to his left.

This flanking action by the squad met with an immediate and violent counter-attack by an estimated enemy platoon. The squad members fought gallantly but soon were overrun. All but one man was killed. The sole survivor, badly wounded, wisely feigned death and later escaped.

The volume and din of enemy fire to the immediate front of the Third Platoon intensified to the point that communications became almost impossible. Lieutenant Heaney continued to try to gain the initiative in his sector, but each move cost him casualties and failed to diminish the enemy fire. Then the enemy began thrusting at the flanks of the company column.

One enemy automatic rifleman penetrated to within 15 feet of the company command post, located immediately behind the Third Platoon, killing the forward observer's radio operator and wounding two others of the group before a grenade from a rifleman of the Second Platoon dispatched him.

Coleman deployed the Second Platoon to both flanks of the company, linking with the now recoiling elements of the Third. One squad, moving to the right, ran into a hail of automatic weapons fire and took several casualties. The squad counterattacked twice trying to reach its dead, but was beaten back into a position that eventually became part of the company's perimeter.

At about this time (1420), the weather closed in and a torrential downpour drenched the battlefield. The 200-foot high jungle canopy admitted limited light at best, and with the storm, only an eerie twilight penetrated to the jungle floor. The deteriorating weather also precluded the airlanding of any immediate reinforcements for the company. The number of casualties taken by the company, plus the violence of the enemy assaults, led Coleman to decide to abandon efforts to regain the offensive, and concentrate on forming a defensible position on the small patch of high ground the company held.

The First Platoon, next in column, was deployed to the right and rear, assuming control of the one squad of the Second. The Weapons Platoon was assigned to fill the remaining gap. The Weapons Platoon, fighting as riflemen, had barely closed into the perimeter and established some semblance of fighting positions across the trail leading into the position, when the enemy launched a determined attack from the west. It was beaten off, but it served notice to the company that it was cut off and facing an enemy force of much greater strength. Moreover, the enemy dead within the perimeter were clad in khaki uniforms and had the helmet normally associated with North Vietnamese regulars or Viet Cong main force units.

By 1630 hours, Bravo had tightened its perimeter and conducted two probes to the east to recover wounded and dead. The wounded were collected at a central location in the perimeter and the unit medics consolidated to treat them.

The enemy forces also recognized that it had isolated the company and began pressing attacks at various portions of the perimeter, preceding each with a heavy concentration of rockets and grenade launcher fire. Each attack was driven off, but at a cost in dead and wounded that was beginning to sap the vitality of the company.

The heavy rains continued unabated during the early evening. Because of the steep terrain tube artillery was unable to place their volleys near enough to the perimeter to be effective in the close-in defense fights. Under the severe weather conditions, the aerial rocket artillery, the usual savior of an isolated airmobile element, could not be expected to function. However, two birds from the 2d Battalion, 20th Artillery (the battalion commander and his XO volunteered for the mission) felt their way up the side of the mountain and, guided by radio, pressed home salvo after salvo of rockets, some within a few feet of the company's perimeter. One last volley, in conjunction with a violent exchange of small arms and automatic weapons fire, ended the enemy threat for the night. The attacks diminished in strength and intensity and by 2000 hours, all contact was broken.

With more than 40 wounded men, along with nearly 20 killed, Coleman had little choice but to remain in position and await reinforcements which had been landed at landing zone HEREFORD during a brief period when the weather broke. The men of Bravo used the respite to continue digging in, and the medics, three of them now badly wounded, worked unceasingly to give aid and comfort to the wounded.

Even though the company was badly hurt, and the men were well aware of their danger the unit maintained its fighting spirit. Not one trooper suggested that the unit pull out. Many indicated their willingness to stand and die on that piece of ground.

At a few minutes after 2200 hours, Bravo with less than 45 effectives actually manning the perimeter was reinforced by some 130 men of Company A, 1st Battalion (Airborne), 12th Cavalry. Alfa, commanded by Captain John W. ("Jackie") Cummings, had air assaulted into landing zone HEREFORD and marched without opposition to the perimeter.

To facilitate relief in the pitch blackness, elements of the relieving company were fed into the line and integrated into overall defense. In many cases, members of Bravo Company gave up entirely or shared their hard dug holes with members of the fresh unit.

On 17 May, at 0615 hours, Bravo pulled in its listening posts, which had been stationed some 20–25 meters outside the perimeter, and the two companies initiated a "mad minute" of fire—a systematic spraying of trees and bushes in front of the positions. This firing touched off an immediate enemy reaction, and he launced a violent attack at all sectors of the perimeter, covering his assaults with an intense barrage of grenade and rocket launcher projectiles. The intensity and violence of the incoming fire indicated an assault by at least a battalion-sized unit.

Both companies fought bravely side-by-side for nearly two hours. Enemy riflemen came within a few feet of foxhole positions before being killed, and the ammunition in the perimeter began running alarmingly low. The approach of another relief company—Company C, 1st Battalion, 12th Cavalry—moving up from HEREFORD, apparently caused the enemy to break contact. As the enemy riflemen faded back into the jungle, the men in the perimeter already had fixed bayonets and had loaded their last magazines in their rifles.

When the smoke cleared, casualties were counted. Bravo had lost 25 killed, and 62 wounded. Alfa had 3 killed and 37 wounded. There were 38 enemy bodies found within or immediately adjacent to the friendly perimeter. Later evidence indicated that as many as 200 additional enemy had died in the fight. (Long afterward, in upper Binh Dinh Province, a North Vietnamese soldier surrendered to the Division. He said that his battalion had been involved in the battle against B Company and testified that his company had been decimated.)

For Bravo Company, the remainder of 17 May was spent evacuating dead and wounded. For the 1st Cavalry Division, Operation CRAZY HORSE had begun.

The rest of the action took place in the most mountainous and heavily forested area in the province, far from the lowlands. Because of the extraordinarily difficult terrain, aircraft commanders found themselves carrying a maximum load of only two or three soldiers as they went into "elevator shaft" single helicopter landing zones in the triple canopy hilltops, where the aircraft would barely fit in a circle of giant tree trunks. Chinooks hovered over the jungle so that the men could climb down swaying "trooper ladders" through the triple canopy. Nevertheless, in the three weeks of CRAZY HORSE, over 30,000 troops moved by helicopter—an example of the tactical value of airmobility in mountain operations.

The battleground was a complex morass, 3,000 feet from bottom to top and 20 kilometers square. In the fighting it soon became clear that the 2d Viet Cong Regiment was bottled up in these rugged hills, but the Division's companies were having trouble in finding and attacking this elusive enemy. In a new plan, the Division marked off the battle area into pie-shaped sectors and moved the airmobile companies to the outer edges on all sides to set up a double row of ambushes. The artillery then began firing 12,000 to 13,000 rounds per day into the enemy concentrations. The Air Force assisted with tactical strikes and also hit the enemy with B–52 raids almost daily. The North Vietnamese, under this pressure, attempted to escape out of the area and triggered several of the prepared ambushes. The enemy's powerful 2d Regiment was disorganized by heavy losses in these ambushes; the survivors evaded to the north into the An Lao Valley and were not in contact again for several months.

During the final phase of Operation CRAZY HORSE, Republic of Korea forces maintained contact with an enemy battalion for four days, inflicting heavy casualties. In this operation the Division evacuated or destroyed 45 tons of rice, 10 tons of salt, a weapons repair shop, and several large caches of ammunition and medical supplies. Captured documents gave the first indications of the extent of the Viet Cong infrastructure in northeastern Binh Dinh. It was this intelligence that provided targets for ensuing operations.

After CRAZY HORSE, there was a pause in the Battle of Binh Dinh that lasted until early September, while the Division concentrated on battlefields to the west and south. With overlapping operations PAUL REVERE (Pleiku Province), HAWTHORNE (Kontum Province) and NATHAN HALE (Phu Yen Province) taking place many miles apart, the Division's logistics were put to a severe test.

Complicating the 1st Cavalry's maintenance and supply problems was the fact that the Division had far too many nonstandard items in the inventory. This made it almost totally dependent on an unpredictable supply system for repair parts and replacements. There were cases in which items, such as bulldozer parts, were not replaced for more than ten months because they were "special" and not stocked in country. There were several unique supply systems, such as those for aircraft parts and electronic components, operating simultaneously in Vietnam, in addition to the military standard requisitioning and issue procedures.

As for the span of control within an airmobile division, every brigade could handle four maneuver battalions with ease and a fifth battalion with minor communication augmentation at brigade headquarters. The division, in turn, could absorb two additional brigades, for a total of five, provided each of the added brigades brings along its support slice. Therefore, logistics became a restraining factor long before span of control or communications.

CHAPTER VI

Airmobile Developments, 1966

The Genesis of the 1st Aviation Brigade

The Army faced a dilemma at this period of time in that it had spread its aviation assets in Vietnam to support the Army of the Republic of Vietnam divisions, one Republic of Korea division, the 1st U. S. Army Division, the 173d Brigade, and the 1st Brigade of the 101st Airborne with non-organic aviation units. Only the 1st Cavalry Division had sufficient organic aircraft. There was little experience in supporting non-airmobile divisions with separate aviation assets and the Army was not sure of the proper mix.

The finite limitations of available Hueys and Army aviators forced the decision to support the average size division or Vietnamese elements with considerably less than had been devoted to the airmobile division and eventually a ratio was determined of one assault helicopter company per brigade in the II, III, and IV Corps Tactical Zones. At this time, the Marines had the responsibility for aviation support in the I Corps Tactical Zone. But, the mere spreading of limited aviation assets did not solve the problem of organization command and control. It was obvious from the beginning that some sort of centralization was going to be necessary. The companies in Vietnam at that time all had developed distinct methods of operations and procedures which they had worked out on an ad hoc basis with the supported commanders. As a result, it was very difficult to take a company from the highlands and switch its support operations to the Delta or vice versa for it entailed relearning complex command relationships, terrain, and methods of operation. Therefore, a high degree of standardization of training, procedures, and methods of operations was obviously indicated.

In contrast to this need, there was a strong underlying feeling that centralization of Army aviation at a brigade level would be a major step backwards from the fundamental principle that Army aviation had held since World War II, that is decentralization to the lowest possible commander who could use and maintain it.

Furthermore, the ghost of "another Air Corps" kept rising from the graveyard with all the unhappy memories of that early era. There were those who saw the formation of such a brigade as the final positive indication that Army aviation was truly headed for a separate air arm.

The 1st Aviation Brigade was formed provisionally on 1 March 1966 with Brigadier General George P. ("Phip") Seneff in command. General Seneff's career had been sprinkled with key assignments related to airmobility and he had been a major commander in the testing of the 11th Air Assault Division. He was quick to assemble many of his former "11th Air Assault Team" to form a brigade staff and to publish the first handbook and standard operating procedures of the brigade. With the publication of basic operational manuals, the 1st Brigade was able to standardize aviation operational procedures throughout Vietnam, establish training schools, enforce safety regulations, and in general more effectively manage the growing aviation assets.

Essentially, General Seneff commanded the non-organic Army aviation elements in Vietnam, but operational control was vested in the supported ground commander. It was a tribute to the personalities involved that this unusual system worked; retaining the best features of centralized command—training and standardization—and the best features of decentralization—operational control and support. The onus of a large impersonal structure with its inherent unresponsiveness was avoided. Though the ghost of another air corps would not be entirely exorcized for years to come, the 1st Aviation Brigade avoided most of the pitfalls that had been grimly predicted at its onset.

General Seneff, in coordination with Major General William E. DePuy who then commanded the 1st U. S. Division, began a test period to determine the most satisfactory means of supporting a line infantry division with separate aviation elements. While considerable work had been done in this area during 1965, it had grown in different directions depending on the unit supported and the aviation assets available. Using the 11th Combat Aviation Battalion, commanded by Lieutenant Colonel Joseph B. Starker, as the test vehicle, Generals DePuy and Seneff soon ironed out most of the serious problems in matching aviation to the ground units. Because the requirements varied in each part of Vietnam, the brigade tried to collocate one assault helicopter company with each U. S. brigade and in course of time each Republic of Korea brigade. The companies supporting Army of the Republic of Vietnam units were located in the most centralized positions for best support. The

1st Aviation Brigade assigned one combat aviation battalion headquarters in direct support of each infantry division and this battalion headquarters normally worked with that division no matter how many companies might be assigned for a specific mission.

At this time the 52d Aviation Battalion supported the 4th Division in the highlands, the 10th Aviation Battalion supported the brigade of the 101st and the Republic of Korea division, the 11th Combat Aviation Battalion supported the Big Red One, the 214th–the 9th Division, and the 269th–the 25th Division. The 13th Battalion, which was later to become a full group, remained in the Delta. Two aviation groups–the 17th and the 12th–supervised the aviation assets in the II and III Corps Tactical Zones respectively. The aviation group commander was also the aviation officer for the U. S. Field Force commander.

The Caribou Transfer

During the spring of 1966, one of the most emotionally packed debates was reaching its final stages. This would culminate on 6 April 1966 in a formal agreement between the Chief of Staff, U. S. Army and the Chief of Staff, U. S. Air Force to relinquish Army claims to the Caribou and future fixed-wing aircraft designed for tactical airlift. Before discussing this issue, we should put a few basic points into focus.

The keystone to airmobility was–and is–the helicopter and it is easy to forget the versatility and uniqueness of this vehicle. Many futile discussions have been held on the cost, ton-mile capacity, complexity, and limitations of the helicopter when either comparing it to fixed-wing transport or other means of mobility. The simple fact is that no other machine could have possibly accomplished the job of the helicopter. It alone had true vertical capability and could perform those scores of missions ranging from an insertion of a long range patrol to the vertical assault of an entire division; it alone could place artillery on the mountain tops and resupply these isolated bases; it alone could evacuate the wounded out of a chimney landing zone, surrounded by 100 foot trees; it alone could elevate the infantry support weapons and deliver that discreet measure of fire so dear to the survival of a squad. Only the helicopter could place a small bulldozer on a critical piece of terrain or extract another aircraft downed in the jungles and return it to fly again.

Therefore, in discussing the inter-service differences between the Army and the Air Force, one must keep in mind that the helicopter—specifically the tactical transport as represented by the Huey—was the absolute *sine qua non* of the Army's concept of airmobility. General Johnson, then Chief of Staff of the Army, was keenly aware of this basic fact.

At the working level in Vietnam both Army and Air Force officers hammered out practical methods of solving their day-to-day problems and coordinating their efforts. Ever since the conference I mentioned in the first chapter of this study where General White expressed his concern on the size of the Army's fleet, the Air Force had opposed the Army's continued acquisition of the Caribou. By 1966 the Army was operating six Caribou companies in Vietnam. The 7th Air Force in Vietnam, which had been upgraded from the 2d Air Division on 2 April 1966, was severely critical of the Army's method of operation of its Caribou assets. They believed that the Caribou should be incorporated into the Air Force managed Southeast Asia airlift and consequently be more productive from a ton-mile basis.

The Army on its side had some very strong advocates for the Army retaining the Caribou and procuring its turbine powered successor, the Buffalo, as soon as possible. They pointed to the history of the Caribou's responsiveness to the demands of far-flung isolated units and the lack of guarantee that such support would be available under Air Force management. They looked upon the Caribou as a bitterly contested victory for a legitimate Army mission and they were appalled at the mere possibility that the Army might trade this victory for an empty guarantee that they could remain in the helicopter business. It was truly an emotionally charged atmosphere—one that was neither as black nor as white as the critics on both sides appeared to believe.

Prior to the Air Force assumption of the Caribou, most were already under centralized management under Military Assistance Command, Vietnam, and were participating in the Southeast Asia airlift effort. However, one-fourth to one-half of their resources were dedicated to specific Army support. That is, one company had the sole function of supporting the 1st Cavalry Division and others had full-time missions in support of small elements in the Delta and Special Forces. A small portion of the Caribou fleet hauled parts and supplies to isolated Army aviation elements. No doubt some of the Caribou missions were inefficient from a standpoint of payload, but none were considered a luxury by the users.

With this as a background, let's turn to the actual agreement by the two chiefs of staff as signed on 6 April 1966. Because it is important and relatively short, the complete text follows:

The Chief of Staff, United States Army, and the Chief of Staff, United States Air Force, have reached an understanding on the control and employment of certain types of fixed and rotary wing aircraft and are individually and jointly agreed as follows:

A. The Chief of Staff, U. S. Army, agrees to relinquish all claims for CV-2 and CV-7 aircraft and for future fixed wind aircraft designed for tactical airlift. These assets now in the Army inventory will be transferred to the Air Force. (Chief of Staff, Army, and Chief of Staff, Air Force, agree that this does not apply to administrative mission support fixed wing aircraft.)

B. The Chief of Staff, U. S. Air Force, agrees—

(1) To relinquish all claims for helicopters and follow-on rotary wing aircraft which are designed and operated for intra-theater movement, fire support, supply, and resupply of Army forces and those Air Force control elements assigned to DASC and subordinate thereto. (Chief of Staff, Army, and Chief of Staff, Air Force, agree that this does not include rotary wing aircraft employed by Air Force SAW or SAR forces and rotary wing administrative mission support aircraft.) (Chief of Staff, Army, and Chief of Staff, Air Force, agree that the Army and Air Force jointly will continue to develop VTOL aircraft. Dependent upon evolution of this type aircraft, methods of employment and control will be matters for continuing joint consideration by the Army and Air Force.)

(2) That, in cases of operational need, the CV-2, CV-7, and C-123 type aircraft performing supply, resupply, or troop-lift functions in the field army area, may be attached to the subordinate tactical echelons of the field army (corps, division, or subordinate commander), as determined by the appropriate joint/unified commander. (Note: Authority for attachment is established by subsection 6, Sec. 2 of JCS Pub 2, Unified Action Armed Forces (UNAAF).)

(3) To retain the CV-2 and CV-7 aircraft in the Air Force structure and to consult with the Chief of Staff, U. S. Army, prior to changing the force level of, or replacing, these aircraft.

(4) To consult with the Chief of Staff, U. S. Army in order to arrive at takeoff, landing, and load carrying characteristics on follow-on fixed wing aircraft to meet the needs of the Army for supply, resupply, and troop movement functions.

C. The Chief of Staff, U. S. Army, and the Chief of Staff, U. S. Air Force, jointly agree—

(1) To revise all service doctrinal statements, manuals, and other material in variance with the substance and spirit of this agreement.

(2) That the necessary actions resulting from this agreement be completed by 1 January 1967.

You will note that the agreement was far broader than the Caribou problem alone and essentially established without doubt the Army's claim to the helicopter and especially to the armed helicopter. Furthermore, you will note that the future developments of VTOL aircraft were left for future negotiations. Sometimes forgotten, this document established a precedent for attaching Air Force units to ground commanders.

In the context of the times I find it quite easy to understand General Johnson's rationale for signing this controversial agreement. The Army's airmobility program was being threatened by inter-service bickering and budget limitations. There was a distinct possibility that, in the absence of some such agreement, an all-out battle on the very right of the Army to own any aircraft might have occurred. Furthermore, the Army Chief of Staff was a man of uncompromising integrity who had every confidence that both the word and the spirit of this agreement would be carried out by both Services. To their credit, both Services did bend over backwards to make sure that the transfer of the Caribou would be smoothly effected by 1 January 1967. The Aviation School set up special training programs for Air Force pilots and the Army Caribou pilots who were working themselves out of a job could not have been more diligent.

General Johnson closely followed the transfer of the Caribou and its subsequent effectiveness under Air Force management. He charged his staff to keep him informed of progress by frequent analyses. One such study, made after the Air Force had been operating the Caribou for a full year, showed a 12.4 percent increase in hours flown and a 11.4 percent increase in Caribou cargo tonnage. Incidentally, this same study showed a drastic rise in the utilization of the Chinook. Fundamentally, there was no real way to compare the Air Force operations versus the Army. By Air Force standards, Army aviation organizations were woefully under-manned and under-staffed; support and maintenance were accomplished entirely differently by the two Services; and the Air Force had consolidated the Caribou operations in three fixed bases. Much more basic from the ground commanders' viewpoint, the Air Force seemed unwilling to procure more Caribou or its successor—the Buffalo.

In this regard, Lieutenant General Bruce Palmer, Jr., Deputy Commanding General, U. S. Army, Vietnam, on 21 April 1968, wrote General Johnson, the Chief of Staff, as follows:

> . . . Although the Air Force has neither attached the C–7 units to Army supported units nor given the Army operational control, the responsiveness required has been achieved by what is termed dedicated service. At least this is a step in the right direction. However, dedicated service leaves the Air Force with a possible string on the aircraft and units. It will work under conditions such as in the case of our C–7 support where there is strong motivation for all elements of the Air Force to make it work and where there are no competing demands within the Air Force for use of the aircraft. In RVN only one C–7 is dedicated in support of the Air Force; this aircraft provides spare parts support for the C–7 fleet. I question that dedicated service would pro-

vide the same degree of satisfaction in a situation where the Air Force felt that it had authority to override this commitment in order to provide high priority support of Air Force units.

The most frequent complaint received from Army users was the Air Force limitation on crew time. This limitation has in the past prevented completion of some missions. When the extent of the complaint became known to [Brigadier General Burl W. McLaughlin, Commanding General 834th Air Division], he immediately stated that this would be corrected internally in his command. This is a further indication of the current Air Force desire to provide highly satisfactory service.

You will note that the study provides justification for more aircraft and defines in fairly precise terms what a follow-on aircraft should be. The Army must take a firm position on the type of aircraft to be provided in the future if satisfactory service is to continue. Two popular trends in studies and statements of requirements may require considerable Army opposition. The first concerns a VTOL development. Required for the mission is a simple, rugged, easily maintained, dependable VTOL airplane which can be built within today's state-of-the-art (in fact, it exists in the deHaviland Buffalo). Efforts to provide VTOL can only result in either an aircraft that duplicates the helicopter by giving up speed, endurance and low cost, or an aircraft that is very expensive and complex with some severe operational limitations. The second trend relates to size. A comparatively small aircraft is required. The Air Force has always tended to build larger and more powerful aircraft. If the aircraft provided for this mission becomes too large, it will gravitate toward the Common Service Airlift System on the basis of obtaining efficient utilization of its greater lift capacity.

Many studies would follow by both the Air Force and the Army to prove the point, or disprove the point, that the Caribou were better, or worse, under Air Force control. The same arguments on combat effectiveness versus cost effectiveness would surface again, but the transfer of the Caribou must always remain a footnote to the Army's story of airmobility rather than a chapter by itself.

Army Aviation Personnel

During 1966, while I was Commandant of the Army Aviation School, the shortage of Army aviators became critical. This should not have come as a surprise since the Army planners had forecast this shortage for more than a year, but now the buildup of U. S. forces in Vietnam and the concurrent requirement for more airmobile assets made it apparent to commanders at every level.

As early as January 1966, Department of the Army had informed General Westmoreland that all aviator sources had been exhausted and nearly 500 Republic of Vietnam returnees would be needed for additional tours in Vietnam. One hundred and seventy-six of these had less than two years between tours and soon an aviator could only count on twelve months between successive tours in Southeast Asia. Every pilot in the grade of major and below was being assigned to a cockpit position. This made some units extremely "top heavy" in rank and consequently morale suffered. Aviators in a sense felt they were being treated as second class citizens and had serious reservations about their career potential. They alone seemed to be singled out for this repeated duty in Vietnam and it was very hard to explain to their families why this was so.

On 14 June 1966, Department of the Army summed up the seriousness of the pilot shortage. Their projection showed that by the end of Fiscal Year 1966 the Army would have a requirement for approximately 14,300 aviators versus a projected strength of 9,700. The requirements were forecast to grow in fiscal year 1967 from a projection of 16,800 to approximately 21,500. Against this requirement it appeared there would only be a strength of 12,800 aviators. Obviously the real constraint to airmobility growth for some time to come would be the critical shortage of aviators.

It was not until 29 March 1966 that the Army received Secretary of Defense approval to increase its flight training from 120 pilots a month to an active Army output of 410 aviators per month. No matter how fast this training program was accelerated, this training rate could not be reached until April 1967. Nevertheless, the Aviation School did reach a rate of 200 per month in May 1966.

Drastic action had been taken throughout the world to reduce manning levels for all overseas commands except Southeast Asia to an absolute minimum. Aviator strength decreased in Europe to a minimum of 250 and only 34 pilots were available to our forces in Korea. Department of the Army mailed individual letters to almost 2,000 aviators in the Reserves asking for volunteers for active duty, but this effort failed to produce substantial results as only sixty applications were received. The Army wisely decided not to drastically reduce its standards nor dip into the "seed corn" of aviation instructors at the schools. Vietnam would have to live with this shortage and use every internal management tool to make the best use of the assets they had.

When I made my second visit to Vietnam in 1966, I was anxious to glean any information which I could take back to the Army Aviation School to better prepare the students for their service in Vietnam. Fort Rucker at that time was almost exclusively geared and oriented to meet the growing needs in Vietnam and the School made a special effort to debrief all Vietnam returnees in an effort to glean every bit of information that would improve its instruction. The new aviator received 32 weeks of training following a syllabus that used every precious hour to best prepare him for Vietnam. After I became Commandant, it became obvious that the rotary-wing aviator must graduate with more than passing knowledge of instrument flying as most of these students were going directly to combat—without the usual unit training transition which had always been the accepted practice in all Services. We managed to squeeze sufficient hours out of the course to give each initial-entry student a thorough instrument course, complete in every respect except for flight checks in two systems not being used in Vietnam—instrument landing system and VHF omnidirectional range. Upon graduation, these students were perfectly capable of flying instruments; but, any instrument course demands continued practice to develop confidence and increase proficiency. Unfortunately many graduates soon lost this capability through lack of use and supervision.

To partially compensate for the lack of unit training after graduation, the course ended with an extended field maneuver. I had directed that this last week before graduation incorporate as much combat realism as possible to facilitate the essential transition from "student" to "pilot." Almost every commander was enthusiastic over the continued high quality of the new pilots being assigned to Vietnam.

If there was ever a question about the wisdom of the Army's Warrant Officer Aviator Program, it was laid to rest at this time. As Commandant of the Aviation School, I watched thousands of these splendid young men mature in front of my instructors into professionals—and that is exactly what they wanted to become—the best Army helicopter pilots in the business. Nothing more and nothing less.

When the Army Deputy Chief of Staff for Personnel and his Assistant visited the Aviation School, they were concerned about the lack of branch career schools for the warrant aviator and his future motivation. They could not believe that most of these people came into the Army for the single purpose of flying; and that, though many had been to college and were married, they did not aspire to a career as a commissioned officer. Further, they were not aware that those few who did want a commission could get one under current regulations. I did my best to disabuse them of their misconceptions. Our warrant program was attracting the very best of American youth at the most critical time.

To add to the problems brought about by the chronic pilot shortage, the Army in Vietnam was given the additional missions of training U. S. Naval aviators in the armed Huey to take over the responsibility for "Market Time" operations. For some time the U. S. Army armed helicopters had been giving fire support to Task Force 116 in their mission of waterway and off-shore surveillance. Now the Navy wanted to train its own pilots and borrow some of the Army's precious armed helicopters to do a Navy mission that no fixed-wing aircraft could do.

On 25 July 1966 a joint U. S. Army–U. S. Air Force–Vietnamese Air Force conference laid down the plans for the Army to train the first ten Vietnamese pilots in the UH–1 helicopter. This was the beginning of a long range plan to expand the capabilities of the Vietnamese to conduct airmobile operations completely on their own.

Finally, the 1st Aviation Brigade was given the mission to familiarize a limited number of Korean Army pilots and mechanics with the UH–1 helicopter. When you add these requirements to the requirement to conduct an in-country familiarization course for all new Army aviators as well as the day-to-day commitments to actual combat, it is easy to appreciate the dilemma of the senior aviators in spreading their thin assets.

In most battalions, each new aviator was given a check ride to ascertain his knowledge of the unit aircraft and, if necessary, was given additional training in this particular type. He was then placed in the copilot seat regardless of his rank and received a theater procedural orientation flying administrative type missions for 25 hours. After this step, he was allowed to fly copilot during actual combat assaults and, when the aircraft commander felt he had sufficiently demonstrated his proficiency, only then was he allowed to fly as first pilot.

Because the Army aviator was anxious to perform as many possible tasks as he could, it became common in the Republic of Vietnam for aviators to fly over 100 hours a month and 120 hours was not exceptional. This amount far exceeded the limit that had commonly been accepted as safe over an extended period of time. Fatigue was inevitably linked to a higher accident rate and commanders at all echelons were alerted to detect signs of pilot fatigue within their own units. The 1st Aviation Brigade established a policy on aviator fatigue which provided close supervision by a flight surgeon of any aviator who exceeded 90 hours of flight time in a consecutive 30 day period. Naturally, flying hours alone were an inadequate measure of this problem: one must consider the intangibles of landing in a hot landing zone versus a "milk run" to Saigon; or the differences between landing time after time in a pinnacle approach versus ordinary operations.

Although regulations required two pilots per aircraft for each flight, some commanders found it necessary to relax this procedure for certain administrative flights. Consequently, many units began an unofficial training program whereby crew chiefs were given "stick" time and were coached in the rudiments of flying to the point where they could take over in case of an emergency and theoretically land the helicopter safely. This training paid off in isolated cases where the pilot and copilot were wounded and the crew chief brought the aircraft back. This informal training had an ancillary benefit in that it encouraged quite a few crew chiefs to apply for warrant officer flight training.

The shortage of pilots was only part of the personnel problem. Skilled mechanics, crew chiefs, avionics and armament specialists were in critical demand. The twelve month rotational policy (which General Westmoreland determined essential for the long range haul) generated special problems in what was known as the "hard skills." The enlisted student might spend a full year or more in developing proficiency in repairing radar or turbine engines, and then serve in Vietnam for a year or less before he had to be replaced by an equally skilled man. For the career soldier, there was a danger that he could become over-specialized in some particular skill required in Southeast Asia and consequently could not be utilized when he left. This led to morale and promotion problems. Many a skilled helicopter crew chief upon being reassigned to the United States or Europe found himself "technically unemployed" and wound up in an obscure motor pool where his special identity would get lost. Fortunately, many of these outstanding people were kept in the program and volunteered either to return to Vietnam for a successive tour or volunteered for flight training.

The performance of the crew chiefs and mechanics were summed up by one general this way:

> The super performance did not stop with the pilots. The maintainers achieved the miraculous. High birds in the brigade were getting a 140 to 150 hours a month piled on them when they were programed for only 70 and the average UH-1D was going over 100. This overflying was in direct conflict with the desired availability rate of 75 percent, but our maintenance detachments met both goals. Crew chiefs flew all day and worked on their birds all night. The sight around the average company maintenance detachment when the birds staggered home in the evenings was a sight to behold. The maintenance crews rolled out, turned on the lights, worked with flashlights, worked by feel, worked any way, in the rain, in high winds and dust storms, all

night long if necessary to patch up the aircraft, pull the required inspections, correct deficiencies and get them back on the line by the next morning. Night test flights, which are prohibited under peacetime conditions, were the rule rather than the exception.

Units in Vietnam would continue to be undermanned throughout 1966 and the ratio of aviation units to ground units would grow increasingly worse for the next eighteen months. Though every effort was being made in the United States to meet the surge demands for pilots and aircraft, there was only so much that could be done to increase the output of the schools and speed up production lines.

"Arc Light"

The 6th of July 1966 marked the beginning of a new close support capability for the ground commander. On this date, the B-52's operated for the first time using a "combat sky spot" bombing system. This system enabled a ground radar control to direct the bomber over the target and also indicate the exact moment of bomb release in almost any kind of weather. Now the big bombers could be used on targets of opportunity with a great deal of flexibility.

The B-52's were not new to Vietnam and, in fact, had made their first strike in War Zone D more than a year before on 18 June. Since that time they had flown more than 3,700 sorties using the code name "Arc Light." B-52's stationed on Guam were being used increasingly on enemy supply routes and suspected bases that were located by the combined intelligence effort of the U. S. and Vietnamese forces.

The new system employing ground radar control also incorporated a quick reaction force of six B-52's which were on continuous alert at Guam. General Westmoreland, commenting on their effectiveness, said, "The B-52's were so valuable that I personally dealt with requests from field commanders, reviewed the targets, and normally allocated the available bomber resources on a daily basis. I also continued to urge that action be taken to substantially increase the B-52 sorties."

To some the use of a heavy bomber, designed for strategic nuclear response, in a counter insurgency environment was analogous to using a sledge hammer to swat flies. To the contrary, I think the employment of this system showed the imagination and flexibility of the U. S. Military in adapting this powerful system to add a new dimension of fire support for the ground trooper. In contrast to the snide remarks of "jungle agriculture" that some armchair strategists applied to the B–52 effectiveness, those who had witnessed this effect on the ground will testify that the B–52 was an awesome weapon which could destroy the deepest Viet Cong tunnel structure, open up instant landing zones, and strike terror into the hearts of the enemy. A single B–52 usually carried 36,000 pounds of "iron bombs." Like any new system, it took some time to smoothly integrate it into the overall fire plan and develop confidence in its control. However, many a prisoner of war substantiated the psychological effect of the B–52's and many enemy dead gave mute evidence of its lethality.

The B–52's relation to the story of airmobility is in the firepower half of the firepower–maneuver equation. The best utilization of the B–52 power included an immediate follow up of a strike, and the air assault was the natural means of such exploitation. In the next few years this combination would be used more and more.

Techniques of the 101st

A civilian might be surprised to learn that the most important training takes place in combat. This would hardly surprise the veterans of World War II and Korea. Every theater of war—and each area of that theater—presents special problems that require the commander on the scene to develop new tactics and techniques to fit the situation. The helicopter was probably the only solution to the dense tropical jungle, but even this versatile machine needed some place to touch down. Every unit in Vietnam had to adapt many of its airmobile procedures to fit its mission. The following sample, from the 1966 files of the 1st Brigade, 101st Airborne Division, is typical.

Brigadier General Willard Pearson, Commanding General, 1st Brigade, 101st Airborne Division, had established a training program for helicopter rappelling techniques since his brigade frequently operated in dense jungle terrain which did not have accessible landing zones. This training was particularly concentrated in the reconnaissance elements and the engineer landing zone clearing teams. These latter teams performed a most necessary and dangerous task of going into an unknown and lightly protected area—with equipment that had to be airdropped or sling delivered—and felling enough trees to permit several helicopters to land simultaneously. Vietnam abounds with many large hardwood forests which are extremely difficult to cut, even with the best heavy equipment. To add to the frustration of clearing operations, the engineer teams found that chain saws were of little value in bamboo forests. Vines became easily entangled in the saws and the bamboo splinters caused many lacerations among the workers.

A corollary problem in the dense jungle was the evacuation of wounded. The Huey at this time had no suitable hoist which would allow a stretcher to be lifted to a hovering helicopter. The hoist of the Chinook was adequate, but too slow for personnel evacuation, and seldom would this helicopter be available for this mission. General Pearson recommended that improved winch systems for both the Huey and the Chinook be provided as soon as possible.

In the 101st operations in the highlands during this period, an airmobile company was placed in direct support of each infantry battalion and the same company habitually supported a specific battalion. The brigade found this arrangement was mutually advantageous; resulted in increased responsiveness; and enhanced the effectiveness of aviation support. By now the use of a command and control helicopter had become routine for each infantry battalion commander and he used this helicopter for liaison, communications relay with subordinate units, assisting units to pinpoint their locations, guiding units to terrain objectives, and locating potential landing zones.

Seldom mentioned in dispatches, the brigade had two fixed wing Beavers which handled a multitude of minor resupply and administrative missions for the brigade; and the old reliable fixed wing O–1 observation aircraft did yeoman service. These airplanes, which were almost taken for granted because they required so little attention, relieved critical helicopter assets for vital tactical missions for which they would otherwise have not been available.

Airmobility and the U. S. "Presence"

With the arrival of the 196th Light Infantry Brigade on 14 August 1966, the total U. S. strength in Vietnam rose to approximately 300,000. More troops would soon follow. To some casual observers throughout the world, this seemed an inordinate number of military personnel to cope with an unsophisticated enemy in a relatively small country. Some people visualized a U. S. soldier on every other square yard of South Vietnamese soil. This simply is not so.

Map 3 is an outline map of South Vietnam superimposed over an outline map of the United States. You will note that when the northern border along the 17th Parallel is positioned near Pitts-

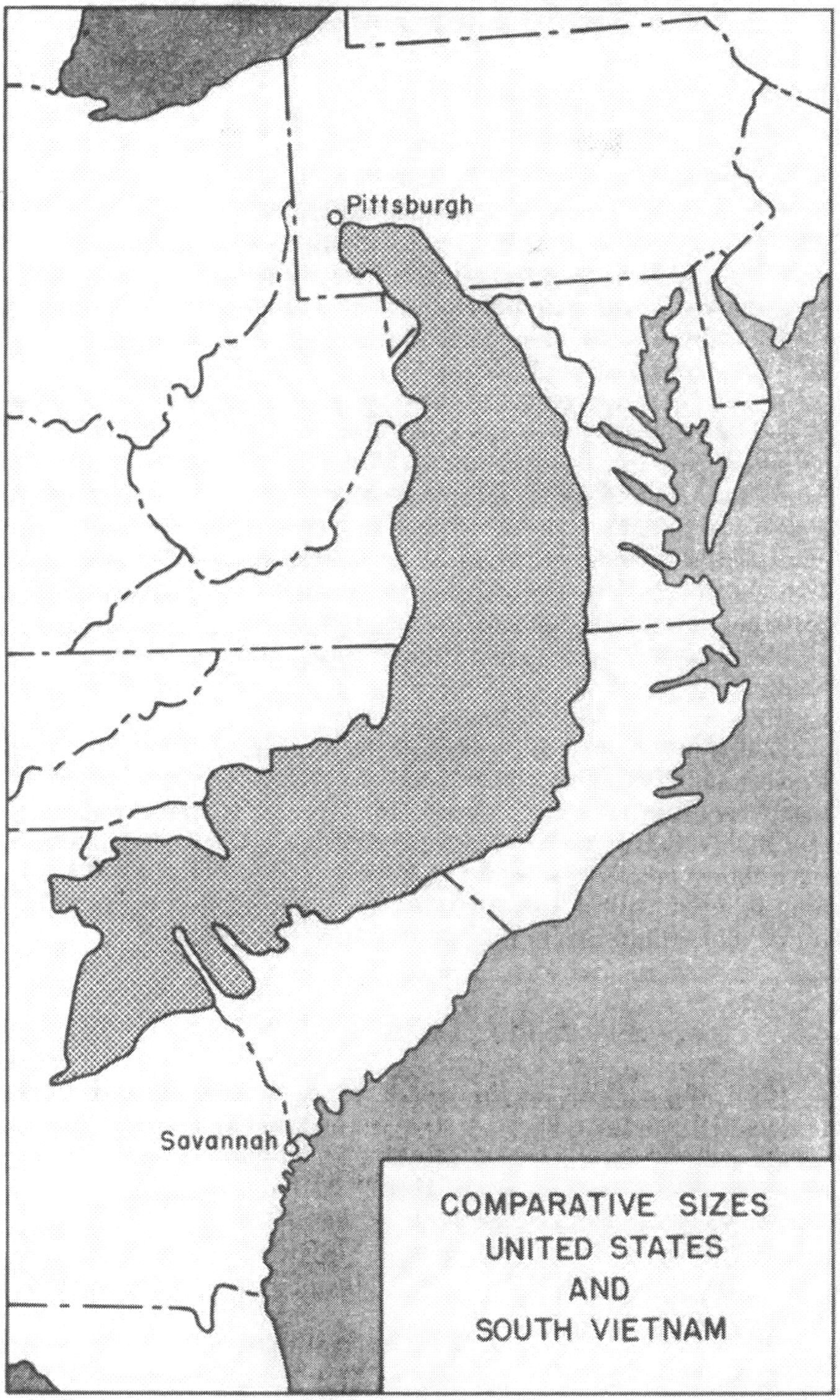
Pittsburgh
Savannah
COMPARATIVE SIZES
UNITED STATES
AND
SOUTH VIETNAM

MAP 3

burg, the southernmost point in South Vietnam (Ca Mau Peninsula) extends to Savannah.

The country is relatively narrow, to be sure, but the long border between Laos and Cambodia is largely a tortuous terrain of jungles and mountains. Even the primitive nomadic tribes, which live in this sparsely populated area, have difficulty in traversing this area once they have left their own little familiar area.

The U. S. forces occupied small dots (not vast areas) on this long expanse. They extended their power by major airmobile sweeps throughout their assigned area of operations, but in no sense did they "hold" the terrain for long periods of time. Many operations inflicted severe damage on the Viet Cong and North Vietnamese troops, but the tenacious infrastructure of the guerrilla in many areas allowed them to rebuild their strength, sometimes in a matter of months. It was often necessary to go back into an area time and time again to defeat not the same enemy but perhaps the same numbered unit which had regrouped from local recruits and replacements from the north. This was particularly true along the borders of Cambodia and Laos in those provinces where a strong communist infrastructure had existed for more than a decade.

Without the potential for extending our power through the helicopter, we would have been forced into small enclaves which in themselves would be prime targets, such as the French had found themselves in at an earlier time. But, air assault techniques gave the initiative to the Free World Forces along with the elements of surprise and mass. No matter what the frustrations were in fighting in Vietnam, it is safe to say that without the airmobile tactics our so-called "massive presence" in Vietnam to a great extent would have turned into mere pinpoints of static defense.

Fall, 1966

Operation THAYER I marked the beginning of the series of battles that kept the 1st Cavalry Division in constant operation in the plains of Binh Dinh for many months. The course of this battle followed the enemy as he drifted across 506 Valley into the Crescent Plains and Cay Giep Mountains.

The Division jumped off in the attack on 15 September 1966, with the simultaneous air assault of two brigades. Three battalions were lifted from An Khe and two from Hammond into the mountains of the Kim Son Valley. The five assaulting battalions secured the high ground all the way around the claw-shaped valley and then fought their way down to the valley floor against elements of the

18th North Vietnamese Army Regiment. In the action that followed, the 3d Brigade was committed, putting a full airmobile division into combat on a single tactical operation for the first time. Later in the fight, the cavalrymen were reinforced by the 3d Brigade of the 25th Division, which increased to eleven the number of battalions controlled by the 1st Cavalry Division, operating in conjunction with nine battalions of the 22d Army of the Republic of Vietnam Division and two to three of Government of Vietnam Marines and airborne troops. The division supported all of these units with its own organic airlift.

On 20 September, the battle area shifted to 506 Valley as the 18th North Vietnamese Army Regiment attempted to evade to the east and break contact. Three cavalry battalions made air assaults to the east to follow their trail. The brigade fire base at Hammond was attacked on the night of 23 September, in what was apparently a move to take the pressure off the enemy in the Kim Son and 506 Valleys as they moved eastward. Also on 23 September, the Capitol Republic of Korea Infantry Division moved into the Phu Cat mountains in force, opening up a new phase of the Free World effort in Binh Dinh Province. A few days later, the build-up of sightings and small actions were clearly indicating that the bulk of the 18th Regiment was shifting toward the coast actually moving into a natural pocket bounded by sea and mountains. In the face of the Government of Vietnam successes in reestablishing governmental control of the coastal area east of Phu My-Phu Cat, the 18th Regiment was being sent in among the population in force to bolster resistence. THAYER came to a close at the end of September with over 200 enemy killed and 100 tons of rice captured.

By the 1st of October 1966, the 1st Cavalry Division had been in Vietnam for a full year and many airmobile techniques had been refined and polished in the crucible of combat. Nevertheless, it is surprising how well the basic fundamentals of the organization stood the test of time.

At the end of its first year of operations in Vietnam the 1st Cavalry Division was to face another major challenge—rotation. During the summer of 1966 approximately 9,000 officers and men of the Division had rotated to the United States. Replacement personnel were flown to Pleiku, South Vietnam, from Travis Air Force Base, California. Liaison teams were placed at both Travis and Pleiku and the massive aerial replacement service from the United States direct to the 1st Cavalry Division functioned with speed and efficiency. Of particular value was the liaison team placed at Travis Air Force Base. This team not only looked after the welfare of

CH–54 Skyhook Helicopter Delivering 155-mm Howitzer

incoming replacements, but coordinated directly with the Air Force troop carrier unit on aircraft, weight, and space available and loading and departure times.

At the other end of the pipeline, at Pleiku, Division personnel received and coordinated the further movement of replacement personnel to the division base at An Khe. Some of this movement was accomplished by convoy over Highway 19, but a significant part of the movement was made by air, as Air Force C–130 aircraft and the Division's Caribou (CV–2) aircraft were available. Although 11,000 miles long, stretching half-way around the world, the Division's replacement pipeline worked smoothly and effectively. Units were maintained at fighting strength, operations continued, and the Division moved through its transitional "changing of the guard."

This changeover not only included the loss of combat seasoned leaders, fighters, flyers, and supporters, but also witnessed the departure of those personnel who had been a part of the birth of the airmobile experience, had been with it through its early development and formative period, had tested it, and had applied the principles of airmobility to the nature of warfare in South Vietnam.

Artillery in the Airmobile Concept

I have mentioned earlier that one cannot view the airmobile concept as a simple problem of moving men and equipment from point "A" to point "B." One of the fundamental reasons for the development of airmobility was the Army's concern that the balance between firepower and maneuver had swung too far in the direction of firepower. It was envisioned that the helicopter could right this balance. However, with this mobility, there was also danger that the soldier would outrun his supporting firepower if some means were not found to increase the mobility of the supporting weapons to the same degree. This is particularly true of artillery.

The U.S. infantryman over the years has come to expect and get continuous artillery support on call. The artilleryman in turn has depended on the infantry to secure his positions and keep his supply routes open. In Vietnam no simple solutions were available to continue this long-established teamwork. The early designers of the airmobile division had recognized that they would have to sacrifice the heavy 155-mm howitzers and be content with moving the 105-mm howitzer with the Chinook helicopter. As it turned out, a 155-mm howitzer battalion was continuously attached to the 1st Cavalry. It was teamed with the CH–54 Crane to become an integral part of the Division's fire support, even though it was not officially "organic." The Little John rocket had been included in the original organization, but when the 1st Cavalry deployed to Vietnam, the Little John was deleted due to tactical and manpower considerations. To make up for this deficiency in firepower, an aerial artillery battalion was organized as the general support artillery. It consisted of three firing batteries, each equipped with twelve Huey helicopters armed with 2.75-inch aerial rockets.

The airmobile tactics of the 1st Cavalry Division, its speed of maneuver, and the distances involved required drastic changes in the techniques and development of fire support co-ordination. For one thing, the air was filled with a number of new objects—hundreds of troop transport helicopters, armed helicopters, reconnaisance aircraft, and tactical air support. Through this same atmosphere thousands of shells from tube artillery had to travel. Fire support co-ordination during the critical air assault phase of an operation was the most difficult to resolve. The tactical air support, tube and aerial artillery, and sometimes naval gunfire and B–52 bombers had to be integrated without danger to the friendly forces and without firepower gaps that would relieve the pressure on the enemy. Only careful planning and carefully worked out standing operating procedures could make this manageable.

As an example, the 1st Cavalry Division had a zone system based on the twelve hour clock superimposed on the map location of each firing position. North was at 12:00 o'clock. Prior to firing, the artillery units announced over the aircraft guard frequencies the danger areas such as "firing in zones three and four, altitude 3,000 feet." It was incumbent on the pilot to check for artillery fire prior to approaching a landing zone. As an additional precaution, canonneers scanned the skies in the direction of fire prior to pulling the lanyard while observers checked the forward end of the trajectory.

During the first battle of the Ia Drang Valley, the 1st Cavalry had covered such a wide zone that the placement of artillery was of utmost importance. Not only was it necessary to cover the landing zones of the attacking forces, but it was also necessary to place artillery in such a position that it could cover another artillery unit for mutual support. In this "war without front lines," the artilleryman found himself often confronting the enemy face to face. It was not long before the artillery decided it would be good practice to collocate itself with the infantry battalion reserve. Nonetheless, many an artilleryman thought he had well deserved the Combat Infantry Badge.

From the artillery viewpoint, the most significant development of the first year's operation of the 1st Cavalry Division was that aerial artillery came of age. In the beginning, some officers considered aerial rocket artillery to be a nuisance on the battlefield, and they could not understand why its method of employment should be any different than the armed helicopter which escorted the tactical troop lift. Fortunately, the aggressive spirit of the aerial rocket artillery battery demonstrated that it could and should be used in mass. While the other armed helicopters were preoccupied by pinpoint targets, the aerial rocket artillery could deliver *area* fire with the same responsiveness (and through the same communication channels) that tube artillery had done in the past.

These same aerial rocket artillery assets, if distributed piecemeal throughout the division, could not have functioned nearly as well. In this case, the whole was greater than the sum of all its parts. Aerial rocket artillery never entirely replaced tube artillery, nor was it ever meant to. However, by careful allocation of fire and precise timing of ships on station, the aerial rocket artillery was able to provide a remarkable volume of fire at times when no other fire support was available. To their credit, the aerial rocket artillery developed the same combined arms partnership with the infantryman that has always existed between the red leg and the dog soldier.

To the infantryman who has never been in an airmobile division, it is a very difficult thing to explain the responsiveness and effectiveness of aerial rocket arillery. Aerial rocket artillery, in contrast to the roving gunship, generally stayed on the ground with one section having a two-minute alert time. They were positioned so to be only a matter of minutes away from any potential target. Two minutes after a fire mission was received they were airborne, and, in a matter of a few minutes more, on target. As soon as the first section departed, a second section moved from a five-minute alert to the two-minute alert status, and another standby team moved to the five-minute alert. By this method, continuous and accurate firepower could be delivered. Aerial rocket artillery was so effective in the 1st Cavalry Division that the artillery commanders had to constantly remind the Infantry to use tube artillery when appropriate rather than call automatically for aerial rocket artillery support.

During MASHER–WHITE WING, the 155-mm howitzer was airlifted for the first time using the CH–54 "flying crane" helicopter. During this same campaign, it became an accepted technique to select hilltops for artillery positions since these were easier to defend and provided open fields of fire.

Artillery has always been notorious for consuming large tonnages of ammunition. Again this is symptomatic with the kind of support that the U. S. infantryman has taken for granted. However, in Vietnam where practically every round had to be delivered by air, artillery ammunition proved to be one of the biggest logistics problems. Commanders had to exert supervision at all levels to make sure that the right fire support means was chosen for the target of the moment. A wise commander did not spend too much ammunition on harassment and interdiction fires that could not be observed.

The airmobile artilleryman had learned to fire in all directions with a minimum of confusion. He became accustomed to rapid and frequent moves and developed confidence in his airmobile prime mover, the Chinook. He even developed his own special tactics, known as "the artillery raid." In this case, an artillery battery would be moved deep into suspected enemy territory, rapidly fire prepared concentrations on targets that had been developed by intelligence, and then pull out before the enemy could react. Some of these raids were conceived, planned, and executed in less than three hours.

Workable methods were found to employ aerial rocket artillery at night, but the mainstay of fire support during the hours of darkness remained the tube artillery. Seldom was an infantry unit required to hold a position at night out of range of its friendly guns.

Close-in fire support has always been inherently dangerous. In the fluid situation of the airmobile battlefield, the ever present danger of the proverbial "short round" is multiplied by all the points on the compass. An error in any direction may well result in friendly casualties. As a result, coordination and control of all fire, the knowledge of the exact location of every friendly element is more important in the airmobile division than in any other combat force. A year's experience in Vietnam had matured the 1st Cavalry Division's fire support techniques and had proven its organization to be fundamentally sound. The artilleryman had not been left behind in this new dimension of combat.

Other Operations

The fall of 1966 saw many operations develop as the U. S. strength continued to grow. The newly arrived 196th Light Infantry Brigade ran into a major enemy force south of Sui Da while searching for rice and other enemy supplies on 19 October. When four companies of the U. S. 5th Special Forces Group's Mobile Strike Force were inserted into landing zones north and east of Sui Da, they immediately became heavily engaged. It became apparent that the Viet Cong 9th Division, consisting of three regiments, together with the North Vietnamese 101st Regiment had deployed into the central Tay Ninh Province with the major objective of wiping out the Special Forces camp at Sui Da. The four Special Forces companies were overrun and had to withdraw in small groups or be extracted by helicopters.

General Westmoreland responded to this large enemy threat by commiting the 1st Infantry Division, contingents of the 4th and 25th Infantry Divisions, and the 173d Airborne Brigade. Some 22,000 U. S. and Allied troops were committed to the battle which became known as Operation ATTLEBORO. The battle continued until 24 November, during which over 1,100 enemy were killed and huge quantities of weapons, ammunition, and supplies were captured. The Viet Cong 9th Division would not be seen again until the following year.

The 1st Cavalry Division continued its operations in Binh Dinh Province with Operations THAYER II and IRVING. During the latter operation on 15 October, they found a medical cache with several thousand containers of medicine and, at another location, camera equipment with 5,000 reels of film. Among the cameras found was one owned by *Look* Magazine senior editor, Sam Casten, who had been killed during Operation CRAZY HORSE in May. Over 1,170 prisoners were captured and identified as Viet Cong infrastructure members or North Vietnamese regulars during Operation IRVING. This operation included the hard fought battle of Phu Huu, in which the 1st Battalion, 12th Cavalry, won the Presidential Unit Citation.

IRVING was a good example of cooperation between U.S., Government of Vietnam, and Republic of Korea forces. The tactical moves and the complicated phasing were carried out with precision, and without any major difficulties. At the same time, the operation showed that large scale sweeps, with two or three divisions participating, should be the exception rather than the rule in this kind of counterinsurgency operational environment. They serve to stir up the enemy, and they bring some of the infrastructure to the surface, but larger units are able to evade the sweeping and blocking forces. In this case, the primary target was the infrastructure within the population, and the sweep was profitable, but, with the exception of the fight at Phu Huu, the North Vietnamese Army forces known to be in the objective area were able to exfiltrate.

During the same period, PAUL REVERE IV was continuing near the Cambodian border in Pleiku Province. The newly arrived 4th Infantry Division carried the bulk of this battle with elements of the 25th Infantry Division and 1st Cavalry Division. By 30 December, 977 enemy had been killed.

Each one of the above operations contained hundreds of examples of the growing capability of U. S. forces to employ airmobile operations with effectiveness and daring. The year 1966 was the year of accelerated buildup and the beginning of major offensive operations in Vietnam. It was also the year in which the enemy began to doubt his strategy and his tactics.

By the end of 1966, the United States would have a total of 385,000 U. S. military personnel in South Vietnam and would be in a position for the first time to go over to the offensive on a broad and sustained basis. General Westmoreland remarked, "During 1966, airmobile operations came of age. All maneuver battalions became skilled in the use of the helicopter for tactical transportation to achieve surprise and out-maneuver the enemy."

CHAPTER VII

The Peak Year, 1967

Parachute Assault in Vietnam

At 0900 hours on 22 February 1967, Brigadier General John R. Deane, Jr., stood in the door of a C–130 aircraft. When the green light flashed, General Deane jumped, leading the first U. S. parachute assault in the Republic of Vietnam, and the first such assault since the Korean conflict fifteen years earlier. This parachute jump of the 2d Battalion, 503d Infantry, signalled the beginning of Operation JUNCTION CITY ALTERNATE. The original plan, as conceived in November 1966, called for the 1st Brigade of the 101st Airborne Division to make the parachute assault; but, much to their chagrin, they were engaged in other operations and the honor was to go to the 173d.

Operation JUNCTION CITY employed the 1st and 25th Infantry Divisions, the 11th Armored Cavalry Regiment, the 196th Light Infantry Brigade, elements of the 4th and 9th Infantry Divisions, and South Vietnamese units, as well as the 173d Airborne Brigade. Their target was enemy bases north of Tay Ninh City, in the area the French had named "War Zone C." The decision to make a paratroop assault was based on the urgency to place a large force on the ground as quickly as possible and still have enough helicopter assets to make a sizeable heliborne assault as an immediate follow-up.

The requirement for helicopter lift on D–day was substantial. The 1st Infantry Division had five infantry battalions to put in by air assault and the 173d had three infantry battalions. In addition to the requirement for the Huey slicks, there was a tremendous requirement for CH–47 lift for positioning artillery and resupply of ammunition. The 173d had computed that they would free 60 Hueys and six Chinooks for support of other forces by using the parachute assault technique. The paratroopers were assigned landing zones farthest to the north—areas that would have cost many extra minutes of flying time for lift helicopters. The practical aspects of making more helicopters available were perhaps colored by the emotional and psychological motives of this proud unit which was anxious to prove the value of the parachute badge; nevertheless, the jumpers contributed strongly to the overall attack.

The 173d was placed under the operational control of the 1st Infantry Division for this operation and developed an elaborate deception plan to avoid possible compromise of the drop zone. In the planning phase only the commanding general, his deputy, and two key staff officers were aware of the actual drop zone. The cover plan designated a larger alternate drop zone outside the planned area of operation. This permitted all the necessary staging preparations which must precede an air drop and all necessary coordination with the Air Force. The actual drop plan for the airborne assault phase of the operation was not distributed to the units until 1900 hours on 21 February, the evening before D–day. After Lieutenant Colonel Robert H. Sigholtz, the Airborne Task Force Commander, briefed his troops on the operation, he sealed off his battalion area as a security measure. Thirteen C–130's were used for the personnel drop and eight C–130's for heavy drop of equipment. Jump altitude was 1,000 feet.

The battalion dropped on schedule and by 0920 hours on D–day all companies were in their locations around the drop zone. Out of the 780 combat troops who made the assault, only eleven sustained minor injuries. The heavy equipment drop commenced at 0925 hours and continued throughout the day. The 1st Battalion, 503d Infantry began landing by helicopter assault at 1035 hours and the entire battalion was in place shortly thereafter. No direct contact with an enemy force occurred during these early hours of D–day. Another infantry battalion, the 4th Battalion, 503d Infantry, conducted a heliborne assault into two other close landing zones at 1420 hours and phase one of JUNCTION CITY ALTERNATE was essentially complete.

During this operation, the 173d Brigade was supported by the 11th, the 145th, and the 1st Aviation Battalions. Over 9,700 sorties were flown in support of the operation and Army aviation lifted 9,518 troops and a daily average of fifty tons of cargo. While the initial parachute assault phase received most of the publicity, the subsequent tactical moves were made by helicopter and the momentum of the operation depended on this support.

As might be expected, some operational problems resulted from this first mix of parachute and heliborne operations. One accident and several near accidents were experienced as a result of helicopters trying to land in an area littered with parachutes. There just wasn't time to adequately police the drop zone. Also there were some problems on Tactical Air coordination. Long delays were imposed prior to a tactical air strike by the communications and coordination required between the large number of units involved. While safety was of prime consideration, there were periods of time when no ordnance was being delivered against the enemy. In addition, General Deane had noticed that air strikes were being called in when troops were unable to break contact, forcing the jets to break away without having an alternate target. However, the control and coordination procedures began to smooth out after the first few hours of confusion.

This combined operation, which the 173d Brigade had begun so dramatically, continued until mid-May. The enemy lost over 2,700 dead along with vast amounts of ammunition, medical supplies, and more than 800 tons of rice. War Zone C, which had been an exclusive Viet Cong stronghold for many years, was now vulnerable to the allied forces at any time of their choosing. In retrospect, there is no question that the parachute assault which began JUNCTION CITY ALTERNATE was effective. The troopers had been well trained and knew what to expect but, as General Deane stated, "More importantly, they did what was expected of them."

The employment of the airborne parachute force is historically visualized as a theater-controlled operation aimed at achieving strategic surprise. Although parachute delivery of troops and equipment is a relatively inefficient means of introduction into combat, the very existence of this capability complicates the enemy's planning and offers the friendly commander one more option of surprise. In this instance, which involved multiple units in a major operation, there was a greater demand for helicopter lift than there were helicopter assets. As a result, the 173d pushed strongly for a parachute assault. The fact that airborne techniques were not used more often in Vietnam can be attributed to many factors.

The most obvious restraint to an airborne operation in Vietnam was the time lag inherent in airborne operations in responding to intelligence on the elusive enemy. The relatively unsuccessful French parachute operations already had pointed this out to us. A much more important restraint was the nature of the war itself and the limitations imposed on U. S. forces. From a strategic point of view, the U. S. posture in Vietnam was defensive. U. S. tactical offensive operations were limited to the confines of South Vietnam. Had the rules been changed, the parachute potential could have profitably been employed by planning an airborne assault into enemy territory at a distance within the ferry range of the Huey.

This would have allowed the parachute force to secure a landing zone and construct a hasty airstrip. Fixed-wing aircraft would have air-dropped or airlanded essential fuel and supplies. Then the helicopters could have married up with this force, refueled, and immediately given them tactical mobility out of the airhead. These circumstances never came about.

Every man with jump wings was eager to prove his particular mettle in Vietnam. However, this special talent was not often suited for that enemy, that terrain, and that situation. Nevertheless, I firmly believe that there is a continuing requirement for an airborne capability in the U. S. Army structure.

Change of Command

On 1 April 1967 I assumed command of the 1st Cavalry Division from Major General John Norton. I was fortunate in inheriting an outstanding team of senior commanders. My two assistant division commanders were Brigadier Generals George S. Blanchard, Jr. and Edward De Saussure. My chief of staff was Colonel George W. Casey, and the three brigade commanders were Colonel James C. Smith, to be replaced in a few weeks by Colonel Donald V. ("Snapper") Rattan, Lieutenant Colonel Fred E. Karhohs, and Colonel Jonathan R. Burton. My division artillery was commanded by Colonel George W. Putnam, Jr., and Colonel Howard I. Lukens had the 11th Aviation Group. Colonel Charles D. Daniel had the Support Command, to be shortly replaced by Colonel Hubert S. ("Bill") Campbell. These officers, along with many others too numerous to mention, enhanced the pleasure of commanding the "First Team." For the next nine months my tactical command post would be in Binh Dinh Province at a location with the unpretentious name of Landing Zone Two Bits. Division Rear would remain at An Khe.

In preparing this section of the study, there was a temptation to view all problems from the vantage point of the division commander. The 1st Cavalry Division was not unique in having many incidental day-to-day problems—problems that were typical of every division in Vietnam. For example, as any commander, I was concerned that we averaged over fifty cases of malaria per month. Again, the 1st Cavalry is fundamentally an infantry division and we had all the problems of booby traps, mines, ambushes, and base security that one would find with any division, *plus* those additional problems that go with taking care of over 400 helicopters. I have tried to focus on the airmobility aspects of my experience in Vietnam, since all other details are covered in the specific history of the 1st Cavalry Division.

The 1st Cavalry Division had been operating in the Binh Dinh Province through four successive campaigns since early 1966—Operations THAYER I, IRVING, THAYER II, and the then current campaign, PERSHING. The 1st Cavalry had put intense pressure on the North Vietnamese Army 3d Division and its three main force regiments throughout these campaigns. They had suffered severe logistical and personnel losses. Relentlessly pursued in every portion of the province, the remaining North Vietnamese Army units were forced to find an area in which they could regain strength and reorganize their ranks. Consequently, the 2d Viet Cong Regiment as early as mid-October 1966 had retreated north into Quang Ngai Province to avoid the 1st Cavalry. The 22d North Vietnamese Army Regiment also withdrew into Quang Ngai during the month of March. These frequent enemy retreats to the north, to rest and regroup, contributed to the necessity for the 1st Cavalry's participation in Operation LEJEUNE which began on 7 April. The principal reason behind this operation was an urgent Marine requirement to free some of their troops in Quang Ngai for movement further north.

Operation Lejeune

The boundary between the provinces of Quang Ngai to the north and Binh Dinh to the south established the demarcation line between the I Corps and II Corps Tactical Zones. This same boundary line divided the U. S. military effort, with the III Marine Amphibious Force having the responsibility in the I Corps area.

Throughout its previous nineteen months of operations in Vietnam, the 1st Cavalry had never operated outside of the II Corps area and, as a matter of fact, no U.S. Army combat unit had operated in the I Corps Tactical Zone. The enemy (with some reason) had assumed that they would be secure by moving into the "safe" area of Quang Ngai Province and thus elude the seemingly endless pursuit by the 1st Cavalry Division. This proved to be a mistaken assumption for on the 7th of April an Air Cavalry battalion task force moved into Duc Pho in the southernmost district of Quang Ngai.

The Duc Pho area had been effectively controlled by the communists for more than ten years. Over the years the Viet Cong and its political arm, the National Liberation Front, had increased their power by political indoctrination, torture, and terrorism until it had a well-developed infrastructure from the sea coast as far inland as Ba To. The legitimate government of Vietnam controlled at most 10 percent of the land area in the district and, in essence, was powerless in the area.

On 28 January, the 3d Battalion, 7th Marines, part of Task Force X-RAY,[1] had moved from its base at Chu Lai into the Duc Pho area to alleviate some of the enemy control in that district. The Marines concentrated their operations around the city of Duc Pho itself, for the Marine battalion had very little helicopter support and was, by and large, restricted to ground mobility. The only Marine reinforcement in this area had been in late February and early March when a special landing force of the 1st Battalion, 4th Marines (from Okinawa) and the 2d battalion, 5th Marines (from Chu Lai) were engaged in a search and clear mission to the south of Duc Pho City. In early March these reinforcements returned to their home bases leaving the 3d battalion, 7th Marines in place. The III Marine Amphibious Force had been receiving increasing pressure along the DMZ throughout the early months of 1967 and increased enemy infiltration had forced the Marines to commit the majority of their forces along this line. The Marines were thin on the DMZ and wanted to move Task Force X-RAY north from the southern part of their area of responsibility. The Marine plan included pulling out of Duc Pho as soon as Military Assistance Command, Vietnam, could provide a replacement force. General Westmoreland decided that the most responsive unit available would be the 1st Cavalry Division.

The 1st Cavalry Division was given less than twelve hours to put a battalion task force into the Duc Pho area and less than 36 hours to increase that force to brigade size. In deference to the Marines, the operation was named after Major General John Archer Lejeune, a Marine leader during the Spanish American War and World War I. Most of the landing zones also were given names from Marine Corps history.

When the orders to move were received on the night of 6 April, the 2d Battalion, 5th Cavalry, under the command of Lieutenant Colonel Robert D. Stevenson, was chosen to begin the action since this battalion was in a preparatory posture at An Khe for another operation. At 0115 hours on the 7th, approximately three hours after the Division had received the oral order, the 11th Aviation

---

[1] Task Force X-RAY was commanded by Brigadier General Foster C. ("Frosty") LaHue. He and I would be coordinating on many operations in I Corps Tactical Zone during the next few months.

Group, located at landing zone TWO BITS, was alerted to move the 2d battalion, 5th Cavalry and supporting elements from An Khe to Duc Pho beginning at first light. The operation officially began at 0930 hours on the 7th. The battalion had fully closed by 1700 hours that afternoon.

It was immediately obvious that the first requirement in this area would be the building of a heavy duty airstrip for support by Air Force aircraft. The decision was made to build a C-7A Caribou strip immediately at landing zone MONTEZUMA which could be expanded to accommodate C-123 aircraft. At landing zone MONTEZUMA there would also be space enough to build a parallel Caribou strip while the first airstrip was improved and surfaced to handle the larger and heavier C-130 aircraft.

Company B of the 8th Engineer Battalion had arrived at landing zone MONTEZUMA during the morning of the 7th and immediately began a thorough reconnaissance of the airfield site. During the next two days, 31 pieces of heavy engineer equipment weighing over 200 tons were airlifted into Duc Pho. This move required 29 CH-54 "Flying Crane" sorties and 15 Chinook sorties. Much of the equipment had to be partially disassembled to reduce the weight to a transportable helicopter load. By 1800 hours on the 7th, enough equipment was on the ground to begin work. The earthmoving commenced and continued throughout the night by floodlights. By midnight, six hours after construction had begun, 25 percent of the Caribou strip was completed.

On 8 April the remainder of the 2d Brigade Task Force under Lieutenant Colonel Karhohs deployed into the LEJEUNE area of operations and assumed operational control of the area. The remainder of the Marine task force was placed under operational control of the 2d Brigade until such time as they could be moved north.

Landing zone MONTEZUMA was composed of light sandy soil; and the heavy rotary wing traffic soon generated monumental, semipermanent dust clouds. This in turn generated a severe maintenance problem in the rotor heads of the helicopters. Peneprine, an oil-base dust palliative, was spread on the helipads and refueling areas as fast as it became available and helped reduce this problem. A concurrent problem was the air traffic control necessary for the hundreds of aircraft arriving with the Air Cavalry combat and combat support units being lifted into MONTEZUMA throughout the day and the Marine aircraft beginning the outward deployment. Air traffic at the landing zone was at best confusing and at worst downright hazardous. As had been experienced in every landing zone (even when there was no confusion), the turbulence from hovering helicopters was severe. Tents were blown down and equipment scattered. One irate individual anonymously broadcast over the air to whom it may concern, "If you blow my tent down one more time, I'm going to shoot you down." A three-man pathfinder team from the Pathfinder Platoon of the 11th General Support Aviation Company soon arrived to assist the Marine air traffic control personnel. The control problem improved, but it was far from being solved.

I had sent one of my assistant division commanders, General Blanchard, to Duc Pho as my personal representative and he remained in the LEJEUNE area of operations almost exclusively. On several occasions, he had expressed to me his deep personal concern over the aircraft density but, nevertheless, was amazed at the effectiveness of the three-man pathfinder team which controlled daily the more than 1,000 arrivals and departures of OH–13, UH–1B, UH–1D, CH–47, and CH–54 helicopters along with the fixed-wing aircraft. The team initially stood on the hood of a ¼-ton truck in the dust storms churned up by rotor turbulence and with a single radio controlled traffic with nearly the efficiency of an air-conditioned tower at a large U. S. airport.

By 1630 on the afternoon of 8 April, the 1500-foot Caribou strip was completed. Work continued on the strip to expand it to 2,300 feet for C–123 use. The first Caribou airplanes that landed carried a mundane cargo of 30 tons of culvert, which was unloaded by the side of the runway since no parking ramp had yet been prepared. Work continued through that night again with glaring searchlights. The Marines shuddered at the Army's intense illumination since they had long been accustomed to very strict light and noise control at night. Their apprehension was fortunately unfounded and the engineer work continued without enemy interruption. On 13 April, the 8th Engineers began the construction of a second Caribou airstrip parallel and west of the completed C–123 strip. This was necessary to allow the C–123 field to be improved and surfaced to meet C–130 criterion. The second strip was finished in 25 hours after 4150 cubic yards of earth had been moved and graded.

One reason that landing zone MONTEZUMA did not sustain an enemy night attack during Operation LEJEUNE was the Night Hunter operations that were conducted almost every evening. Karhohs developed this technique to a high degree in this action and, later, on the coastal plains of Binh Dinh. These operations used four helicopters; one acted as a lead and flare ship, while the other three unlighted helicopters were armed to take care of targets of opportunity. As the lead ship dropped flares, the door gunners in the next two helicopters, which flew at a higher altitude and at a distance not to be revealed by the flares, observed the ground with starlight scopes. Once the enemy was spotted, the gunners would open up with tracers to pinpoint the target for the last armed helicopter to make a run with 2.75-inch rockets. This technique was very effective at finding and killing the enemy and denying him one of his most valuable assets—the night.

Once the brigade task force had established itself at Duc Pho, a sea line of communication became necessary to provide the required daily 250 short tons of supplies necessary to support the brigade task force. The newly constructed Caribou airstrip could not possibly carry all the required logistical support. A sea terminal point was selected almost directly east of landing zone MONTEZUMA called Razor Back Beach. During the 8-day period between 9 and 17 April, over 8,000 short tons were moved over the shore from LST's and LCM's. The operation was not without difficulty. Sand bars prevented LST's from coming in with a full load and eight to twelve foot breakers made the landings of the smaller craft extremely touchy.

Operation LEJEUNE combined the efforts of four military services: The U. S. Army, the U. S. Marine Corps, the U. S. Air Force, and the U. S. Navy. The tactical air force support was substantial. They had only taken over operation of the Caribou since the first of that year and, though slightly leary of accepting the hastily constructed airstrip, they made 159 sorties into MONTEZUMA carrying 1,081 passengers and 229 tons of cargo. During the first eight days of operations, tactical air dropped 115 tons of bombs and 70 tons of napalm. The U. S. Navy provided gun support with a total of 2,348 rounds from its two ships offshore, the USS *Picking* and the USS *Shelton*.

It was apparent right from the beginning that the enemy in that area had never before been challenged by airmobile tactics. It was several days before they began to appreciate the versatility and flexibility of the 1st Cavalry operations. The enemy chose to disperse and hide. Contact was only sporadic with the heaviest combat action on 16 April.

At noon on 22 April, Operation LEJEUNE was terminated. Although contacts were primarily light throughout the operation, 176 enemy had been killed and 127 captured by the 2d Brigade, 1st Cavalry. This brigade was relieved in place by the 3d Brigade

COMBAT ASSAULT—TROOPS MOVING OUT TO SECURE THE LANDING ZONE

of the 25th Division, under Colonel James G. Shanahan, which had been under the operational control of the 1st Cavalry Division for some months. The 3d Brigade of the 25th would soon become part of Task Force OREGON, the first division-size U. S. Army force to be formed in the I Corps Tactical Zone.

Operation LEJEUNE was unique in many ways. The deployment of the 2d Brigade to the I Corps Tactical Zone was the first commitment of any large U. S. Army unit in that area. More importantly, the engineering effort, including the lifting of 30 tons of equipment to build two tactical fixed-wing airstrips in a matter of a few days, was unparalleled in Army engineering history. Finally, the demonstrated "fire brigade" reaction capability of deploying a large task force in a day and a half to an entirely new area of operations proved again the flexibility of the airmobile division. At Duc Pho the 1st Cavalry left behind two airstrips, an impressive sea line of communications, several critical connecting roads, and a damaged Viet Cong infrastructure. In light of the limited mission, Operation LEJEUNE was an unqualified success.[2]

The techniques of the 1st Cavalry Division that were demonstrated in Operation LEJEUNE were to prove invaluable in the many campaigns to follow. Operation LEJEUNE, however, was a relatively short move outside of the Division's area of operations. Much more complex and longer moves were made shortly thereafter which involved Air Force fixed-wing aircraft in addition to the organic lift. These moves demonstrated a technique whereby an Air Cavalry unit was extracted from combat, moved to a landing zone, and the bulk of its equipment flown by Air Force aircraft to a new location. The organic aircraft would then be ferried to the area and join with the unit ready to be employed again.

The Cavalry Spread Thin

On 23 June 1967, about 0900 hours in the morning, Lieutenant General Stanley R. Larsen told me to have a battalion ready to move by 1300 that afternoon. They would be lifted by C–130's from landing zone ENGLISH in the Bong Son Plain to the Dak To–Kontum area in the Central Highlands where they were desperately needed. It was necessary to pull the 2d Battalion, 12th Cavalry directly out of contact with the enemy and get them to landing zone ENGLISH. Using 24 C–130 aircraft loads and two C–123's, the battalion moved to Dak To by that evening along with an artillery battery. They were almost immediately thrown into combat. The next day, two more battalions followed and the third day, the remainder of the direct support artillery. By now I had my 3d Brigade (–) committed to operational control of the 4th Infantry Division. In the next few days these units would participate in one of the hardest fought battles of the Vietnam war. The enemy had shown unexpected strength and determination. During this action the 3d Brigade was commanded by Colonel James O. McKenna, who had just taken over the Brigade from Colonel Burton on 22 June. The 3d Brigade would not return to my control until 25 July.

---

[2] Operation LEJEUNE had an interesting sidelight in its unique command relationships. I mentioned earlier that we had crossed the invisible border into I Corps Tactical Zone where Lieutenant General Lewis W. Walt, as Commanding General, III Marine Amphibious Force, was commander of all U.S. forces in that area. On the other hand, I worked for Lieutenant General Stanley R. Larsen, Commanding General, I Field Force, Vietnam, who commanded all the U.S. forces in II Corps Tactical Zone. (Are you still with me?) Even though I commanded a major operation on General Walt's real estate, I reported directly to General Larsen and only co-ordinated informally with the Marines on the spot. This is hardly standard military procedure.

In order to cover the area left open in the PERSHING area of operations, I spread the remainder of the 1st Cavalry Division north and south of the Bong Son River. With the Division's unique ability to rapidly generate a reserve force from other forces in contact, I had no hesitation in allowing all maneuver elements to be committed.

This is not to say that any commander is happy to lose operational control of any of his forces. The 2d Battalion, 7th Infantry that had been detached from the 1st Cavalry before I took command had been operating in the southeast corner of the II Corps Tactical Zone near Phan Thiet since September 1966 and circumstances dictated that I would not have it back under my wing until after January 1968, when it would join me in the I Corps area. Administratively, it still belonged to the Division and we were responsible for all normal support for this unit except operational control, which was vested directly to General Larsen.

Throughout the Battle of Binh Dinh, one airmobile battalion task force was detached under I Field Force control in Binh Thuan Province to support pacification activities around the city of Phan Thiet. The task force was created and moved on 24 hours notice. Although scheduled for 60 days of operation, it stayed in being for 17 months. The task force contained a very significant part of the division's assets. In addition to the 2d Battalion, 7th Cavalry, it included a scout section from the air cavalry squadron, a platoon of engineers, a battery of 105-mm howitzers, a platoon of aerial rocket artillery, lift helicopters, a signal team, and intelligence and civil affairs personnel, plus a forward support element for logistics. This battalion-sized operation, known as Operation BYRD, was especially interesting as a parallel to the Division's activities during this period. BYRD was in effect a microcosm of the Division's operations in Binh Dinh.[3]

Binh Thuan is located about 100 miles northeast of Saigon, 200 miles south of Binh Dinh and bordering the South China Sea. The principal port city, Phan Thiet, was surrounded by a heavily popu-

---

[3] Task Force BYRD was originally under the command of Lieutenant Colonel Billy M. Vaughn. It was in turn commanded by Lieutenant Colonel Fred E. Karhohs who made several significant changes in the Task Force's operations. Upon General Norton's recommendation, Karhohs took over the 2d Brigade the same day I assumed command of the Division. Lieutenant Colonel Leo D. Turner became the new commander of Task Force BYRD. During the last six months of the Task Force's operations, it was commanded by Lieutenant Colonel Joseph T. Griffin, Jr., who wrote a definitive thesis on Operation BYRD while attending the U. S. Army War College in 1970.

lated rice-growing area. Forty percent of the Province consisted of forested mountains, which supplied some of the best timber in Vietnam. These woodlands also provided clandestine bases and rest areas for the Viet Cong. Operation Byrd was an economy of force effort, using a minimum involvement of United States ground combat forces, aiming to upgrade capabilities of the armed forces of South Vietnam in that area.

In order to protect the vital port of Phan Thiet and surrounding areas, the Commanding General of I Field Force, Vietnam gave the 1st Cavalry Division the mission of defeating the enemy forces in the Byrd area, in close coordination with South Vietnam forces. The battalion was to assist in opening National Highway #1 as it ran along the coast through this area. Of the approximately 1,600 personnel in the average daily task force strength, 650 were organic to a battalion of the 1st Cavalry Division, 500 were from other 1st Cavalry assets, and 450 were from I Field Force. Although the composition of the task force varied, the nucleus was the airmobile battalion. The small force contained all the elements essential to sustain independent operations and could take advantage of available support.

Initially, the task force established a fire base and command post on the Phan Thiet airfield from which infantry rifle companies were air assaulted into landing zones within the range of the direct support of the artillery battery. The first operations relieved pressure on Phan Thiet and the nearby district capitals. The task force began combined operations with the South Vietnamese, taking advantage of U. S. Navy ships and U. S. Air Force fighters for fire support. The area of influence of the task force was broadened by the establishment of fire bases at steadily increasing distances from Phan Thiet.

The operation, as it continued, isolated the enemy in the heavily-forested areas, away from the populated zones. In the close-in areas, the task force concentrated on the Viet Cong infrastructure and took action to build confidence in the population and in the friendly armed forces. Revolutionary development activities extended from the city toward the outlying parts of the province. All of these phases took place simultaneously. Initially, the task force operated with regional and popular forces of the South Vietnamese. Later, it conducted combined operations with two Vietnamese Army battalions of the 23d Army of the Republic of Vietnam Division.

The Task Force ended up with an amazing record. During the 17 months of Operation Byrd, the 2d Battalion, 7th Cavalry had only 34 troopers killed in action while 849 enemy were killed and 109 captured, More important than enemy losses, the Task Force had enabled the South Vietnamese government to spread its control from the province and district capitals to virtually all the population in the area. Agriculture production, commerce, education and medical treatment had increased manyfold.

The small but vitally important air assaults of Operation BYRD not only reproduced the Binh Dinh battle in miniature; they also underscored the significant advantages of envelopment over penetration as a tactic. Penetration is costly—opening a gap, widening it, and holding the shoulders in order to get into the enemy's rear. The air assault concept permits a cheaper, faster, and more decisive vertical envelopment approach, which has made the conventional battlefield more fluid than ever. The great variety of air assault concepts seems to fall under two major headings—each of which is a principle of war—surprise and security. In the Battle of Binh Dinh as well as in Operation BYRD, extensive preparations or detailed reconnaissance, while maximizing security, compromised surprise and often created dry holes. Executing air assaults without prior artillery preparation and with limited prior reconnaissance involved considerable risk, but frequently yielded rewarding results. The choice depends on the enemy situation and the ability of the G–2 to present the proper recommendation to the division commander. The air assault must rely on speed, scheme of maneuver, locally available firepower (aerial rocket artillery), and command and control from an aerial platform. Additionally, a reinforcing capability to exploit success or to assault the enemy from another direction must be immediately available to the commander.

This period in the BYRD fighting was characterized by almost daily contacts with squad and platoon-size Viet Cong elements as the task force searched base areas and interdicted lines of communication. Combined United States and South Vietnamese operations were continuous, with both sides gaining mutual respect and experience.

Reconnaissance in Force

In August 1967 the 1st Cavalry Division again moved into Quang Ngai Province in I Corps Tactical Zone with three battalions under the 3d Brigade, commanded by Colonel McKenna. This resulted in the first major reconnaissance in force into the

Song Re Valley.[4] The Song Re Valley had been a sacrosanct Viet Cong stronghold for years. The picturesque terrain consisted of numerous hillocks in the valley floor, fertile fields of rice, and well-fed livestock. Previous aerial reconnaissance had drawn heavy antiaircraft fire. Although the valley appeared prosperous, only a few inhabitants had been observed. Intelligence experts suspected that military age personnel were either hiding in the hills as outright Viet Cong soldiers or being used as laborers by the Viet Cong forces in constructing fortified positions.

On 9 August 1967 the 2d Battalion of the 8th Cavalry, under the command of Lieutenant Colonel John E. Stannard, commenced a battalion air assault into the valley 32 miles southwest of Quang Ngai City. The selected landing zone, named landing zone PAT, was situated on a ridgeline 2,300 meters southwest of an abandoned airstrip at Ta Ma. This landing zone selected because it was the only high ground large enough and clear enough of obstructions to allow six lift ships to land, and because it was in an area which would give an assaulting company the advantage of reconnoitering from high ground down to the valley floor. The assault started at 0936 after a short artillery preparation. After the 1st Platoon had landed, intense antiaircraft fire came from the surrounding hills. Two Huey's were shot down almost immediately. Company A of the 2d Battalion was faced with a pitched battle for the next four hours. The enemy situation, reconstructed later from information gained from prisoners of war, captured documents, and a survey of the battle area, disclosed that the chosen landing zone was right in the midst of well-prepared enemy positions. Looking down on the position were at least 80 North Vietnamese with three 12.7-mm antiaircraft weapons, 82-mm mortars, and 57-mm recoilless rifles. A Viet Cong Montagnard rifle company was on the same hill mass. The ridgeline was rimmed with fox holes and well-concealed bunkers almost flush with the ground. Company A had landed in a nest of hornets.

There were to be hundreds of acts of individual heroism in the next few hours which have been duly recorded elsewhere. How-

---

[4] This operation was co-ordinated with the larger operations of Task Force OREGON taking place in southern Quang Ngai Province under Major General William B. Rosson (who in turn reported to III Marine Amphibious Force). Again the corps boundaries had not changed but the 1st Cavalry took its orders from CG, I Field Force Vietnam as it did during Operation LEJEUNE. During this time, the 1st Cavalry worked with the 1st Brigade, 101st, under Brigadier General Salve H. ("Matt") Matheson. Later, when the Brigade joined its parent division, it would be commanded by Colonel John W. ("Rip") Collins.

ever, for the purpose of this study, the importance of landing zone PAT stems from the fact that Air Cavalry units were able to react with terrific firepower and extract their men from almost untenable positions when necessary. The aerial rocket artillery had fired 576 rockets in support of this action and two armed Chinook helicopters had delivered eight tons of ordnance on possible escape routes. Tactical Air had done a magnificent job of supporting the ground forces with a total of 42 sorties. What could have been a disaster turned out to be an effective assault, killing 73 enemy while only losing 11 friendly troops. Two major enemy units had been flushed out of hidden positions and a major antiaircraft position had been destroyed.

The skirmish at landing zone PAT was the major encounter with the Viet Cong during the reconnaissance in force of the Song Re Valley. This reconnaissance was a preview in miniature of major operations of the 1st Cavalry in subsequent years. We had learned to establish a fixed-wing base near the assault area and save our precious helicopter sorties for short-range missions. Song Re Valley was an excellent rehearsal for what was to come.

The Chinook as a "Bomber" and "Flying Tank"

As a commander I could not help but be struck by the never-ending inventiveness of the U. S. soldier. In my experience the average soldier in Vietnam was as good as, and, in many ways, better than, his World War II predecessor. He came up with new solutions for new problems in a new environment almost daily. One example was the use of the Chinook as an "ad hoc bomber."

The Viet Cong had developed tremendous underground fortifications and tunnel systems throughout Binh Dinh Province. Many of these fortifications could withstand almost any explosion. Riot agents were introduced to drive the enemy from his tunnels and force him into the open. During Operation PERSHING the 1st Cavalry dropped a total of 29,600 pounds of these agents from CH–47 aircraft using a simple locally fabricated fusing system on a standard drum. Initially the drums were merely rolled out the back of the open door of the Chinook and the fusing system was armed by a static line which permitted the drum to arm after it was free of the aircraft. Using this method, a large concentration of tear gas could be placed on a suspected area with accuracy.

Napalm was rigged and dropped in a similar manner during this same period. A single CH–47 could drop two and one-half tons of napalm on an enemy installation. Naturally, this method of dropping napalm was only used on specific targets where tactical air could not be effectively used.

Another version of the CH–47 which was unique to the 1st Cavalry Division was the so-called "Go-Go Bird." The "Go-Go Bird," as it was called by the Infantry, was a heavily armed Chinook which the 1st Cavalry Division was asked to test in combat. Three test models were received armed with twin 20-mm Gatling guns, 40-mm grenade launchers, and .50-caliber machine guns, along with other assorted ordnance. Though anything but graceful, it had a tremendous morale effect on the friendly troops which constantly asked for its support.

From the infantryman's viewpoint, when the "Go-Go Bird" came, the enemy disappeared. The pilots who flew these test aircraft performed some incredibly heroic deeds to prove the worth of the machine. However, from the overall viewpoint of the Division, these special machines required an inordinate amount of support and, if we had kept them as part of our formal organization, we would have been required to give up three of our essential lift Chinooks. Army Concept Team in Vietnam monitored the tests of the "Go-Go Birds" and flew many of the missions. After two of the test vehicles were lost through attrition, the final armed Chinook was transferred to the 1st Aviation Brigade. Much debate would continue about the effectiveness and vulnerability of such a large armed helicopter, but the individual trooper who enjoyed its support would never forget it.

Armor in an Airmobile Division?

To deal more effectively on a continuing basis with the enemy fortifications on the coastal plains of Binh Dinh, the Division had requested and received one tank company from the 1st Battalion, 69th Armor, 4th Division, at Pleiku. On the surface one of the most unlikely additions to an airmobile division would appear to be heavy armor. The tank with all its implications of ponderousness seems to be the antithesis of what one looks for in a "lean and mean" light fighting unit. In many circumstances, this would be true. But, I had always wanted to explore the possibility of a combination of the surprise of air cavalry and the shock of armor. Binh Dinh seemed the appropriate place.

The 90-mm gun of the M–48 tank was found to be one of the best weapons in dealing with the enemy fortifications. In 24 to 48 hours, a determined enemy could prepare an elaborate perimeter affording him excellent cover and concealment around an entire village. His fortifications were well organized and usually prepared in a series of hedgerows. The Viet Cong would not usually leave his bunker under any circumstances. The bunker had to be destroyed to kill him. It was here that the tank came into its own.

In the Bong Son Plain, the 1st Cavalry Division usually employed one platoon from A Company of the 1st Battalion, 69th Armor with from six to eight tanks in conjunction with an infantry battalion when they assaulted a fortified village. Because the employment and maneuver of the tanks were so essential in attacking a fortified village, a problem of too much communications sometimes developed. Everybody from the infantry platoon leader and company commander on the ground to the battalion commander and his S–3 in the air were directing and maneuvering the tanks. To sort out this problem, it soon became standard procedure to put the tank commander or his executive officer in a light observation helicopter to control the attached tanks for the infantry battalion commander. This method worked best.

The tank usually carried a basic load of 62 rounds of 90-mm shells and 2,000 rounds of .50 caliber ammunition and very often, in the course of a four or five hour battle, a tank would use as much as three basic loads. To prohibit a tank running out of ammunition during a crucial time of an engagement, the 1st Cavalry developed a system whereby a basic load of tank ammunition was pre-slung for helicopter delivery and stored at the closest fire base to the scene of the action. Upon initial contact, an immediate resupply of ammunition was initiated. This same air transport capability was used to maintain the tanks by rapidly moving mechanics and repair parts to disabled vehicles.

The most critical limiting factor in the use of tanks in Vietnam was the trafficability of the soil. In the Bong Son area during the dry season and the latter stages of the rice-growing cycle, the M–48 tank could move across the rice paddies with a certain amount of ease. When the rice paddies were flooded, movement was greatly restricted and had to be carefully planned in conjunction with the engineers. Bulldozers and engineer mine-sweeping teams had to be attached to the moving tank elements to keep open movement options for the armor.

In September 1967, the 1st Cavalry Division received another armor capability when the 1st Battalion, 50th Mechanized Infantry was attached. The battalion was completely ground mobile in its organic armored personnel carriers.

When I received the 1st Battalion, 50th Mechanized Infantry, I decided not to treat this battalion as an orphan child to be held in reserve for some particular contingency, but rather to totally integrate it into the 1st Cavalry Division and to train its troops completely in airmobile tactics. We rounded out the battalion with a fourth rifle company from headquarters and supply units and placed their armored personnel carriers at a central position near landing zone UPLIFT. The companies would go out on airmobile operations just as other companies of the Division and if a mission appeared that needed a mechanized unit, we extracted the troops to landing zone UPLIFT and deployed them in their primary role. The 1st Battalion, 50th Mechanized Infantry proved to be a very valuable asset and, when we had lost our attached tanks to their parent organization, we often employed the Armored Personnel Carriers with their .50 caliber guns in tank-like formations. In using the mechanized battalion in this manner, we felt we enjoyed the best of both worlds. We had the additional troops which were completely trained in air assault tactics and we had the mechanized capability when the terrain and situation demanded.

The "Cobra" Arrives

On 1 September 1967, the first Huey Cobra (AH-1G) arrived in Vietnam. The initial six aircraft were assigned to their New Equipment Training Team, under the supervision of the 1st Aviation Brigade. Cobra New Equipment Training Team training started on 18 September with pilot transition courses and instruction on air frame, engine, armament and avionics maintenance. The Cobra was a major step forward in the development of the armed helicopter. For those pilots who had been flying the old, make-shift UH-1C's, it was a *giant* step.

After all this time there were many people, both in and out of the military, who didn't understand the role of the armed helicopter. Ever since Colonel Jay D. Vanderpool had tied a machine gun on a H-13 in the mid-50's, there were those who saw the armed helicopter as a fragile toy dreamed up by frustrated fighter pilots in the Army who were unable by regulation and budget to own really sophisticated attack aircraft. The consensus was generally that a semi-skilled skeet shooter or even a good slingshot artist could knock any helicopter out of the sky at short range and that an encounter with more sophisticated antiaircraft weapons would be suicidal. This attitude is quite understandable in duck hunters who never had the challenge of ducks shooting back. Also, the very nature of the helicopter, which looks very ugly and fragile com-

Huey Cobra Firing in Support of a Combat Assault

pared to a sleek jet aircraft, adds to the conviction that flying one in combat is non-habit forming.

On the plus side, the helicopter is the most agile of all aircraft and has a capability of taking advantage of cover and concealment at extremely low altitudes that would be impossible in a fixed-wing airplane. It was soon proven that the helicopter was remarkably hard to shoot down and the most vulnerable part was the pilot himself. Personnel armor protection and armored seats greatly increased the pilot survivability. The experienced pilot used every unique aspect of the helicopter's flight envelope to his advantage.

Observation from the helicopter is unequaled. The enemy learned that to fire at one was to give up his advantage of cover and concealment and generally bring a devastating return of machine gun fire and rockets. A corollary to the advantage of seeing the enemy was the ability to identify our own troops with precision. Consequently, the armed helicopter pilot could safely place fires within a few meters of our own troops. This became particularly important as the enemy developed the "hugging" tactics which he used to avoid the heavier fires from our tactical air support and B–52 bombers.

The Army had long realized that the Huey-gun-rocket combination was a make-shift, albeit, quite ingenious, system that should be replaced by a new aircraft specifically designed for the armed mission. In the early 1960's, industry asserted that advance was within the state of the art. Experts in research and development urged the Army planners to go for a compound helicopter with an integrated armament system as soon as possible. They argued that it was technically feasible and procurement of any "interim" system would mean the Army would be stuck with an inferior capability for years to come. Moreover, it appeared that an advanced system could be procured almost as soon as an interim aircraft.

There were other pressures too. The Office of the Secretary of Defense had been critical of all the Services in their efforts to procure expensive weapon systems that appeared to offer only marginal improvements over the system they were to replace—aircraft that flew a little higher or faster, tanks that had only slightly better performance, ships that cruised but a few knots faster. Ever since the Howze Board, the Army was sensitive to any criticism that it was striving for less than the best in airmobility. Also, the Air Force maintained that much of Army aviation *duplicated* an Air Force capability rather than, as the Army claimed, *complemented* Air Force support. The Army decided that its best option was to hold a design competition for a totally new system that would offer unique capabilities.

Unfortunately, what was a straight forward concept for a new armed helicopter soon became bogged down in a morass of permutations, modifications and additions to its design. The technicians had taken over from the tacticians. The concept grew in complexity and cost. Worse, it was being pushed into a later time frame when it was sorely needed in combat. Such things as a rigid rotor, ground avoidance radar, inertial navigation and computerization were straining the state of the art and pricing the Army out of mass production. A reevaluation was inevitable.

Bell Helicopter Company had prudently carried on its own research and development program using proven dynamic components of the Huey. Consequently, they were able to offer, at the appropriate moment, an "off-the-shelf" armed helicopter for just slightly more than the modified UH–1 that the Army was then buying to replace Vietnam attrition. The "Cobra" had enough speed to meet the escort mission; tandem seating; better armor;

and a better weapons system. With the strong urging of the combat commanders, the Army decided to procure an interim" system for immediate requirements while it sorted out the problems of the "ultimate" system.

DECCA

The history of the use, lack-of-use, and misuse of DECCA has many important lessons for future developers of airmobile equipment.

It was recognized by the earliest planners that one of the limitations of the airmobility concept would be operating at night and under periods of extreme low visibility. Research and Development offered many possible options to improve the helicopter's capability under these conditions, but all were expensive and complex.

The British had perfected a low-level radio navigational aid known as DECCA which essentially used three low frequency ground radio stations to propagate a series of hyperbolic curves which could be translated by a cockpit instrument into a position fix. Accuracy depended on the spread of the stations, the distance from the station, and the weather conditions. Because of the low frequency, one of the attractive features was its low altitude capability. This contrasted to the line-of-sight limitation of omnidirectional radio navigation aid and Tactical Air Navigation used by the Air Force. The Army tested several versions of the DECCA system and decided it had enough advantages to warrant its installation on command helicopters and lead aircraft.

A DECCA chain had been installed in Vietnam and in the early 1960's, the Army took over its supervision and maintenance. A big disadvantage in the DECCA system was the requirement for special maps printed with the hyperbolic grid and a reluctance by the user to take the time and effort to develop confidence in the system. Its use was further complicated by the resistance of the Air Force to accept a position report in instrument weather from a DECCA read-out as a positive fix. Many senior officers were dissatisfied with the accuracy and reliability of the DECCA system at night and eventually the DECCA died from lack of use and misuse. The requirement for a secure, accurate means of low-level navigation remained.

"Fire Brigades" Sent North

On 28 September the 1st Cavalry Division was notified that an increasing enemy buildup in the I Corps Tactical Zone might require that the III Marine Amphibious Force be reinforced. The

1st Cavalry was alerted to prepare one brigade. The brigade began movement on 1 October as the 2d Battalion, 12th Cavalry and the 3d Brigade command post departed for Chu Lai. By that evening the 3d Brigade command post, the 2d Battalion, 12th Cavalry, B Company of the 1st Squadron, 9th Cavalry and C Battery of the 1st Battalion, 77th Artillery were closed and the Brigade was under operational control of the Americal Division.

By 3 October all Brigade elements were in place and Operation WALLOWA began the next day with a series of air assaults. This deployment also involved the first major move of significant maintenance elements from An Khe to Chu Lai. This experience would prove to be extremely valuable in later operations in the I Corps Tactical Zone.

The concept of a brigade task force, in the U. S. Army's current division organization, is such that different battalions can be used under any brigade controlling headquarters. This allows a great deal of flexibility. For example, the 3d Brigade which had been deployed to the Dak To area and returned to the PERSHING area of operations had different battalions when it deployed to Chu Lai. During November we were alerted to move another brigade back to the Kontum-Dak To area. This time I elected that the 1st Brigade headquarters under Colonel Rattan would control the task force.

The second battle of Dak To, under operational control of the 4th Division, has been extensively documented elsewhere. For the purpose of this study, it is important to note that the helicopters of the 4th Infantry Division, the 173d Airborne Brigade, the 52d Aviation Battalion, and the 1st Cavalry Division flew in excess of 10,000 hours in support of the battle. Over 22,000 sorties were flown, transporting 40,000 passengers and 6,000 tons of cargo. Dak To was another example of the flexibility of an airmobile division which allows its assets to be parcelled out as rapid reaction forces and still continues on a basic mission of its own. The large PERSHING area of operation was left with only one thin brigade during this period. I was glad we had spent so much time working with the 22d Army of the Republic of Vietnam Division on airmobile tactics, since the 22d, under the able leadership of Colonel Nguyen Van Hieu, would have to bear the major burden in Binh Dinh Province for a time.

Operation Pershing Continues

During the long period of the Binh Dinh operations, the 1st Cavalry Division had developed a special rapport with the regiments of the 22d Army of the Republic of Vietnam Infantry Division. The Army of the Republic of Vietnam regiments were assigned distinct areas of operation contiguous to the 1st Cavalry brigade areas and, teamed with 1st Cavalry helicopters, they became well versed in the intricacies of airmobile assaults. During Operation PERSHING over 209 joint operations were conducted with the 22d Army of the Republic of Vietnam Infantry Division. The 40th Regiment of this division played a major part in the Battle of Tam Quan.

Back in May 1967, the Division's capabilities had been greatly enhanced by the attachment of three companies from the 816th National Police Field Force. Introduction of the National Police Field Force into the PERSHING area of operations brought a new weapon to bear on the Viet Cong infrastructure. Now, the Division could conduct cordon and search operations of hamlets and villages with greatly increased effectiveness. The National Police Field Force squads were very important to 1st Cavalry operations in the Binh Dinh Province.

Tam Quan

The Battle of Tam Quan, 6 December to 20 December 1967, which was one of the largest battles during Operation PERSHING, was a good example of the "piling on" tactics which had been so successful in the early airmobile reactions to the enemy. The battle began with the fortuitous discovery of an enemy radio antenna by a scout team near the town of Tam Quan and a small force was inserted at 1630 hours on 6 December. Although the original enemy contact had been late in the day, the 1st Brigade reacted by "piling on" with a battalion of infantry and elements of the 1st Battalion, 50th Mechanized Infantry. On the following day, elements of the 40th Army of the Republic of Vietnam Regiment joined the fight and distinguished themselves by their aggressive manner. Throughout the battle, which was characterized by massive use of artillery, tactical air support, and air assaults by both the U.S. and Army of the Republic of Vietnam troops, the allied force held the initiative. There were frequent vicious hand-to-hand battles in the trenches and bunkers. The division used its mechanized forces to fix the enemy and drive him from his fortified positions. The airmobile units hit him when he tried to move. The enemy lost 650 men during this fierce engagement.

The Battle of Tam Quan had a much greater significance than we realized at the time. In that area, it pre-empted the enemy's *Tet* offensive even though the full impact wasn't then realized. As

Awaiting the Second Wave of Combat Helicopters on an Isolated Landing Zone During Operation Pershing

a result, that part of Binh Dinh was the least effected of any part of South Vietnam during *Tet*.

1967 Draws to a Close

During the late fall of 1967, plans had been developed which would have a tremendous effect on the future of the 1st Cavalry operations during the next year. U. S. Army, Vietnam, Headquarters, for the first time in the war, had been given the mission of contingency planning. They began planning four contingencies which would project the 1st Cavalry Division into the I Corps Tactical Zone. I—the Cavalry would go north of Kontum and Pleiku . . . way north of Dak To; II—the Cavalry would go up near Khe Sanh; III—to Ashau; and IV—the Cavalry would go into the big supply area west of Quang Ngai. The plans themselves, known as the York Series, are incidental but, like many contingency plans, they made the U. S. forces examine potential logistical bases, without which the plans were meaningless. The Marines were told to start working on Red Beach north of Da Nang as a logistical base to support the Cavalry in this series of operations. A smaller logistical base was to be set up at Hue-Phu Bai. Events were soon to prove that the logic behind this planning effort by U. S. Army, Vietnam, was indisputably sound.

The year 1967 had proved many important facets of the airmobile concept. Perhaps the most important facet that had been demonstrated without question was the inestimable value of the Air Cavalry squadron. This unit, especially in its operations in the I Corps Tactical Zone, had demonstrated its unique capabilities in uncovering the elusive Viet Cong. Practically every major engagement was started with a contact by the 1st Squadron, 9th Cavalry Troop, and the enemy was very slow in discovering means of coping with this reconnaissance in force.

The Air Cavalry squadron success in the airmobile division convinced higher headquarters that more Air Cavalry squadrons should be assigned to the theater to work with non-airmobile divisions. In my briefings to the many senior officers who visited the Cavalry Division, I never missed an opportunity to state that no matter what kind of a division I might be privileged to command in combat, I would fight tooth and nail for the capability of an Air Cavalry squadron.

CHAPTER VIII

Tet, 1968

Summary of Operation Pershing

Operation PERSHING was officially terminated on 21 January 1968, after almost a full year of fighting in, over and around Binh Dinh Province. To most of us PERSHING had come to mean an area of operation rather than a single campaign. During this time the 1st Cavalry Division had been continually fighting at least two different battles, often more.

The primary battle was the tedious task of routing out the Viet Cong infrastructure—that very real shadow government that had been strong in this area even during the French occupation. Working with the National Police Field Force, the 1st Cavalry participated in over 970 combined operations which had resulted in the identification and removal of over 1,600 members of the Viet Cong political and administrative structure. About 200 Viet Cong were identified as key leaders. Under the Cavalry umbrella, the National Police Field Force searched more than 340,000 individuals and 4,300 dwellings. At the close of PERSHING we felt that 50 percent of the Viet Cong cadre had been rendered ineffective. The pessimist would have to conclude that this left half of the infrastructure intact, but the fact remained that the Government of Vietnam had held an election in this troubled province, wherein 96.9 percent of the people eligible to vote, *voted*. This compared with a nation-wide average of 80.9 percent.

The second continuous battle in Binh Dinh was the regular North Vietnamese Army 3rd Division units. The enemy lost 5,715 soldiers killed in action and 2,323 enemy were captured. Somehow this latter figure seemed to get lost in the statistics, but I always thought it was very important and was a key indicator of the type of operations being fought in this area.

I was often asked if I thought the 1st Cavalry was "wasted or stagnating" in Binh Dinh. In answer, I'd merely show the questioning visitor a map depicting the air assaults of the last month and it immediately became evident that only an airmobile unit could have covered the area. A typical month, October 1967, showed two battalion-size assaults, 110 company and 165 platoon-size assaults. As a matter of fact, the Division was just about as "busy," from an airmobile viewpoint, as it would ever be.

Statistics on relative vulnerability show that out of 1,147 sorties one aircraft would be hit by enemy fire, one aircraft was shot down per 13,461 sorties, and only one aircraft was shot down and lost per *21,194 sorties.* Used properly the helicopter was not the fragile target some doom-forecasters had predicted.

While the two major battles of Operation PERSHING were continuing, the 1st Cavalry Division was the source of "fire brigades" for many other operations in the western and northern provinces of Vietnam. Some of these have been mentioned earlier. On 4 May 1967, the division boundary had actually been moved further north to include a portion of the Nui Sang Mountain and the Nuoc Dinh Valley in I Corps Tactical Zone. Elements of the division had conducted operations in that general area, particularly the An Lao Valley, throughout the remainder of Operation PERSHING.

The words of my third Chief of Staff, Colonel Conrad L. Stansberry, in summing up the after action report of Operation PERSHING bear repeating:

> Operation PERSHING was the largest single operation and the most successful in which the 1st Air Cavalry Division has participated since its arrival in Vietnam in August 1965. While it is difficult to measure fully the degree of success achieved in the many facets of the war in Binh Dinh Province, significant damage was inflicted on the enemy in loss of lives and combat assets. In addition, many gaps in the VC infrastructure were created over a period of time which made operations in the coastal plains area more difficult for him to execute. The local population was made aware of the Free World presence and the ability to continuously defeat and harass the enemy.
>
> There are several reasons why the Division was able to accomplish this. Part of the answer lies in the ability to hop, skip, and jump over the entire AO in short notice; part lies in the close relationship between the 1st Cav Div and the 22d ARVN Div, CRID and other military and paramilitary forces operating in the area; part lies in the fact that the Division operated for a long period of time in the AO and became thoroughly familiar with the terrain and general atmosphere of the area; and part lies in the exploitation of the NPFF in separating the VC infrastructure from the people.
>
> The Air Cavalry Division also has hidden attributes which are not apparent from organization charts, methods of operation, etc., that contributed significantly to the success of Operation PERSHING. Frequent air assaults over the battlefield maintains the "spirit of the offensive" in 1st Cav soldiers and causes them to live with the high degree of flexibility that promotes a "can do" attitude. In addition, the Air

Cav soldiers' identification with a unit of unique capabilities produces a high sense of pride and will to win which was expertly refined by professional leadership at all levels of command. The mere fact that the airmobile soldier arrived rapidly and boldly on the battlefield, fresh and lightly equipped and well supplied contributes significantly to an attitude of willingness to close with and destroy the enemy.

The 1st Air Cavalry Division enjoyed exceptionally high morale during Operation PERSHING, a key factor in successful combat operations. The use of organic helicopters permitted the airmobile soldier to be abundantly supplied with everything he fired in his weapon, drank, ate, and read: even in relatively inaccessible areas on the battlefield. He lived comfortably in the field and traveled and fought equipped with only his weapon, ammunition, and water with other essentials being air-lifted to him when he needed them. The possibility of being wounded did not discourage him because any helicopter on the battlefield whether it be a CC ship, gunship, etc., could and frequently was used for medevac. In addition, the increased firepower, and visible teamwork between organic Air Cav and ground elements never let the soldier feel as if he were alone on the battlefield. A combination of all these factors encouraged a basically tough and well trained soldier to be habitually willing and eager to close with the enemy in order to enjoy the inevitable satisfaction of defeating him.

While the performance of the individual soldier is a key factor in success in combat, a closely related factor is the close teamwork and unity of effort of all elements of the 1st Cav Div. Aviation, maneuver and fire support elements functioned in close harmony and "went the extra mile" in individual effort to achieve success. The performance of all personnel in the division was without exception highly admirable causing the Air Cavalry concept to achieve results beyond expectations in application.

Most of the 1st Cavalry would soon leave Binh Dinh Province to counter an enemy threat to the far north. Operation PERSHING had not by any means wiped out the Viet Cong infrastructure but its success can be summed up in words of an enemy prisoner of war. This man, the Chief of Staff of a sapper battalion, stated: "I do not know whether you have known or not but I can say that during the period from September 1967 to January 1968 the liberating forces were driven near to the abyss, especially in the areas of Khanh Hoa, Phu Yen and Binh Dinh, where ARVN and allied forces enlarged their areas of activities to an extent that the VC had never thought of."

The Enemy Tet Offensive

The time was 0315, 31 January. The tower operator at Tan Son Nhut heliport, Mr. Richard O. Stark, had just received a call from an aircraft requesting to know if the field was secure. He replied in the affirmative. At 0325 the aircraft called again saying he had reports of enemy contacts in the area. Mr. Stark recalls, "I noticed sporadic tracer fire northwest of the helicopter tower, but I was not duly alarmed. Minutes later, when a C–47 departed from Tan Son Nhut and drew heavy ground fire, I realized that this was not nervous guards, but actual enemy contact." Tan Son Nhut Air Base was under attack!

This attack was one of many similar attacks which were launched against military installations and population centers throughout the Republic of Vietnam. These attacks marked an all-out Communist offensive that continued throughout the *Tet* holidays.

Within three minutes after the alert at Tan Son Nhut, two "Razorback" fire teams consisting of four armed helicopters from the 120th Assault Helicopter Company were airborne and attacking the enemy. Major Ronald K. Kollhoff, commander of the 4th Gunship Platoon from this company, said, "The extent of the enemy buildup was surprising. When it first started we expected a small token diversionary force–a suicide squad–to divert attention from an expected mortar attack. But after a while it became evident that the VC wanted to actually take Tan Son Nhut very badly."

Major General Robert R. Williams, Commanding General, 1st Aviation Brigade, and his house quest, Colonel E. Pearce Fleming, Jr., were sleeping in the Long Binh BOQ when the alert sounded there. Within minutes they were in a command Huey checking into the 12th Combat Aviation Group control net. There were ground attacks taking place in several areas of the Long Binh–Bien Hoa perimeter and a number of gunships were already airborne and were being directed to targets. Colonel Fleming reported, "I was impressed with the professionalism of all hands that I observed and heard during this period of the *Tet* offensive. The calmness and the voices of the men on the radio made you think they were merely calling for landing instructions at a peaceful U.S. airfield, and yet they were continuously in action for hours on end. When the sun finally came up on the morning of 31 January, I was surprised to find that the VC were continuing to stay and fight." [1]

---

[1] General Williams, the Army's first Master Aviator, had taken command of the 1st Aviation Brigade from General Seneff on 16 September 1967. Colonel Fleming was on temporary duty in Vietnam from his position as Deputy Director of Army Aviation.

Time magazine, on 9 February 1968, recapitulated the events this way:

Into Saigon in the days just before Tet slipped more than 3,000 Communist soldiers armed with weapons ranging up to machine-gun and bazooka size. Some came openly into the open city, weapons concealed in luggage or under baskets of food, riding buses, taxis and motor scooters, or walking. Others came furtively: some of the Viet Cong who attacked the U. S. embassy had ridden into town concealed in a truckload of flowers. Once in town, they hid their weapons. Only after the attack did Vietnamese intelligence realize that the unusual number of funerals the previous week was no accident: the Viet Cong had buried their weapons in the funeral coffins, dug them up on the night of the assault. They even test-fired their guns during the peak of the Tet celebrations, the sound of shots mingling with that of the firecrackers going off. . . . An enemy force of at least 700 men tackled the city's most vital military target: Tan Son Nhut airstrip and its adjoining MACV compound housing Westmoreland's headquarters and the 7th Air Force Command Center, the nerve centers of U. S. command in the war. The Communists breached the immediate base perimeter, slipping past some 150 outposts without a shot being fired, and got within 1,000 feet of the runways before they were halted in eight hours of bloody hand-to-hand combat. All told, the Communists attacked from 18 different points around Tan Son Nhut, getting close enough to MACV to put bullets through Westy's windows. Westmoreland's staff officers were issued weapons and sent out to help sandbag the compound, and Westmoreland moved into his windowless command room in the center of MACV's first floor. Other Communist units raced through the city shooting at U. S. officers' and enlisted men's billets (BOQ's and BEQ's), Ambassador Ellsworth Bunker's home, Westmoreland's home, the radio and TV stations. Wearing ARVN clothes, raiders seized part of the Vietnamese Joint General Staff Headquarters, turned the defenders' machine guns against helicopters diving in to dislodge them.

There is no doubt that the quick reaction of the armed helicopters saved Tan Son Nhut and Bien Hoa from serious danger of being overrun. In the first few hours they were the only airborne firepower since the Air Force aircraft could not get clearance to even take off. An Air Force sergeant describing the action on a tape recorder at Tan Son Nhut kept repeating over and over, "Oh, those beautiful Huey gunships!" One of the men in those gunships, Captain Chad C. Payne, a fire team leader, said, "I received fire everywhere I turned. My ships received seven hits, but this was nothing considering the amount of ground fire directed toward us. There were hundreds of VC bodies everywhere in the vicinity of the Tan Son Nhut perimeter. I've never seen anything like it."

Another tribute to the effectiveness of the gunships came from a member of Advisory Team 100 at Tan Son Nhut. When he received word that Tan Son Nhut was under attack, he assembled a patrol of 30 men. "And we ran head-on into one of the attack forces. There were approximately 350 men against my 30. We were certainly outnumbered," he said. "Then those beautiful gunships came in and started circling the area. I threw up a pocket flare to mark the position, and the gunship radioed that we were too close to the enemy force and to pull back some, if possible. We pulled back and then he went in. He was right on target, placing rockets right in the middle of Charlie's position. We killed over 200 enemy, and I'd estimate that 80–85 percent was attributable to the helicopters. The morning of the 31st, if I had met that pilot, I'd have kissed him."

Another area of heavy activity was at the U. S. Embassy in downtown Saigon. Chief Warrant Officer Richard Inskeep of the 191st Assault Helicopter Company was the first to land a chopper on the embassy during the heavy fight, bringing ammunition and evacuating one wounded man. "We were receiving fire from all sides," said Mr. Inskeep. "but we couldn't see anybody around so we lifted off. My gunner then spotted someone in a hole of the roof, so we made a tight turn and came back onto the pad. The fire was so intense that the gunner and crew chief had to pull the ammunition out of the ship and crawl across the roof as they pushed it in front of them. They pushed the ammunition down the hole and helped bring the wounded man back across the roof to the ship."

Watching from below was Mr. George Jacobson, Mission Coordinator of the U. S. Embassy. Commenting on the helicopter's approach, Mr. Jacobson said, "He came in low and I thought for a minute he was going to hit the building, but at the last minute he pulled up and made a beautiful landing on the roof. Afterwards I realized that he did it on purpose to avoid the enemy fire. It was a tremendous piece of airmanship." Mr. Jacobson, a retired Army colonel, was to finish off the last guerrilla inside the embassy. As troopers of the 101st were landed on the Embassy's helipad, the enemy guerrilla tried to escape the troopers, spotted Mr. Jacobson, and fired three shots. He missed and Jacobson shot him with a .45 that had quickly been tossed up to his second floor window by troops below. This was the finale to the six and one-half hour battle within the embassy.

As daylight came over the Bien Hoa Air Base, fighting was still raging around the airfield. Small bands of Viet Cong had managed to penetrate the southeast and southwest areas of the air base, and reaction forces were sent out to stop them. The Air Force's 3d Security Squadron found an estimated 100 Viet Cong to be in the southwest area just beyond the taxiway. The Viet Cong were well dug in, and the security force could not flush them out. On finding themselves pinned down, they called on the Cobras of the 334th Armed Helicopter Company to suppress the guerrillas. Air Force Second Lieutenant John A. Novak, who was in command of the security force, said, "As the Cobras came to our support they swept down about two feet over our heads and fired into the enemy position, knocking out the enemy who were pinning us down. I personally witnessed time after time the Cobras sweep into the VC area and pin down the enemy in the face of heavy fire being directed at them. The Cobras were the turning point in the enemy's destruction."

General Williams and Colonel Fleming had made a complete circuit of the III Corps Tactical Zone to determine first hand information on the situation at the various airfields. Colonel Fleming relates:

> In every case we found that everyone had been busy since midnight the night before, both on the ground repelling attacks against their perimeter, and mounting out their gunships, firefly missions, command and control ships, and in many cases conducting air assault operations.
>
> About mid-morning we landed at Bien Hoa and visited the 334th Gun Company and the Cobra NET Team. These guys had had everybody in action since early that morning and perhaps the most spectacular sight was a crippled Cobra approaching from the Saigon area and making a running landing between the revetments at the Bien Hoa heliport.
>
> He had lost his tail rotor drive shaft, had two rounds in the fuel cells, and had a full panel of warning lights, yet he brought it home and stepped out and said, "Put this one aside and give me another one. There are more targets down there."

The Communists had hit in a hundred places from near the Demilitarized Zone in the north all the way to the tiny island of Phu Quoc off the Delta coast some 500 miles to the south. No target was too big or too impossible. In peasant pajamas or openly insigniaed North Vietnamese Army uniforms, the raiders struck at nearly 40 major cities and towns. They attacked 28 of South Vietnam's 44 provincial capitals.

The Tet Offensive at Quang Tri

At 0420 on 31 January, the 812th North Vietnamese Army Regiment and supporting elements launched a concerted attack on the provincial capital of Quang Tri, a key communications hub in the I Corps Tactical Zone. The enemy's timing had been late for a platoon of the 10th Sapper Battalion went into action inside of the city at 0200 hours in the morning and their premature action alerted the city's defense to the impending attack. The blunt of the attack fell upon the defending Army of the Republic of Vietnam forces in and around the city, the 1st Army of the Republic of Vietnam Regiment with two battalions oriented north and northwest of the city. The 9th Army of the Republic of Vietnam Airborne Battalion was located in the suburb of Tri Buu to the east. This battalion became the first troop decisively engaged. Pressure was very heavy on the defending Army of the Republic of Vietnam force and they were gradually forced to fall back into the city, contesting every foot of the ground as they withdrew. By noon on the 31st, the outcome of the battle remained very much in doubt.

Shortly after noon on the 31st, the provincial advisor, Mr. Robert Brewer, held a conference with Colonel Rattan, commander of the 1st Brigade, 1st Cavalry Division, and the U. S. advisor to the 1st Army of the Republic of Vietnam Regiment. Mr. Brewer stated that the situation in the city was desperate and that the enemy had infiltrated at least a battalion inside the thin Army of the Republic of Vietnam lines. It appeared the enemy was reinforcing from the east and had established fire support positions on the eastern and southern fringes of the city. Recognizing the need for immediate action, Colonel Rattan, after a hasty call to division headquarters for authority, agreed to reorient his brigade and attack to the east of the city.

The Move North

I had agreed with Colonel Rattan's assessment of the situation at Quang Tri and I trusted his judgment implicitly. His brigade had arrived in the Quang Tri area on 25 January having moved up from Hue-Phu Bai after initially deploying from Bong Son on 17 January. This move was part of the plan to orient the entire 1st Cavalry Division in the I Corps area. Our initial orientation toward the I Corps area had started with the YORK series of plans which gave us a good head start on base areas at Red Beach and Hue-Phu Bai. The 3d Brigade had been in the I Corps Tactical Zone for some time under the operational control of the Americal Division taking part in Operation WALLAWA between Chu Lai and Da Nang. The 2d Brigade of the 1st Cavalry had remained in the PERSHING area of operations and we were given the newly arrived 2d Brigade of the 101st Airborne Division, under Colonel John H. Cushman.

Our move was far from being complete at the outset of the *Tet* offensive, particularly our support echelon. Throughout the month of February, we would be moving and fighting and at the same time establishing our maintenance and supply bases.

I had already worked out most of the details of moving the Division with Colonel Putnam, my very capable Chief of Staff, in case the order to move was received during my absence. Colonel Stansberry was then head of the Support Command. Between the two of them, they cut through the red tape of inter-service bureaucracy and solved the hour-by-hour problems that arose in moving the division north.

Colonel Putnam had been practicing moving the division command post long before we got the actual alert order to move. Back in Binh Dinh Province, he had held an actual practice move in which the division command post at landing zone Two Bits was moved to An Khe, operated out of An Khe for twenty-four hours, and then transferred back to Two Bits. At that time the heart of the forward command post was contained in two pods carried by the CH–54 Crane. One pod had a G–2, G–3 Operations Center and the other pod contained the Fire Support Coordination Center and other control elements. The forward command post had a complete communications system to pick up operational control of the division as soon as it was in place. Incidentally, these pods which looked very efficient on the surface were really a headache to emplace in Vietnam and terribly immobile once emplaced. After we moved north, we discontinued their use as impractical.

The move north was not without problems. Some of the Marine staff officers were very reluctant to accept Colonel Putnam's estimate on the space required for the division headquarters. Through an old War College friendship with Brigadier General Earl E. Anderson, the Chief of Staff of III Marine Amphibious Force, he was able to get in to see Lieutenant General Robert E. Cushman, Jr., Commanding General, III Marine Amphibious Force, and secure Landing Zone Tombstone, which later became Camp Eagle. Colonel Putnam reported that there was only one bright spot in his entire battle with the staff. The Navy captain who commanded the Seabees came up and said, "Tell me WHAT you want done and in what ORDER you want it done." This was just the first example of the magnificent support we received from the Seabees.

When Colonel Rattan called me his brigade was actually oriented south and west of Quang Tri, with one fire base as far away as 20 kilometers to the west of the city. At that time, he did not have the mission of protecting the city itself. The decision to change the orientation 180 degrees and attack late in the afternoon was a bold one, but was to prove indisputably sound.

The Battle of Quang Tri City

Colonel Rattan had consulted with Mr. Brewer on the most probable location of enemy infiltration routes and then selected landing zones adjacent to them. These assault areas were selected for the purpose of reducing the enemy's reinforcement capability by blocking his avenues of approach and to eliminate his fire support capability by landing on his support areas. At approximately 1345 hours, the 1st Battalion, 12th Cavalry and the 1st Battalion, 5th Cavalry were directed to launch their air assaults as soon as possible with priority of lift assigned to the 1st Battalion, 12th Cavalry. Additional aircraft were requested from division resources, and the 1st Squadron, 9th Cavalry and aerial rocket artillery were alerted.

By 1555, B Company of the 1st Battalion, 12th Cavalry had assaulted into a landing zone east of Quang Tri and a few minutes later was followed by C Company. Even while they were landing, C Company received intense fire from an estimated enemy company. This contact lasted until 1900 at which time the enemy broke contact leaving 29 bodies. The 1st Battalion, 12th Cavalry air assaults had straddled the heavy weapons support of one of the enemy battalions. As a consequence, this battalion found itself heavily engaged on the eastern edge of Quang Tri by the Army of the Republic of Vietnam and in its rear among its support elements by the Air Cavalry troops. Caught between the two forces, it was quickly rendered ineffective. Shortly after the 1st Battalion, 12th Cavalry had launched its attack, the 1st Battalion, 5th Cavalry assaulted southeast of Quang Tri with two companies. They also landed directly on enemy positions and immediately came in heavy contact with the K–6 Battalion of the 812th North Vietnamese Army Regiment. Like its sister battalion, the K–6 found itself wedged between Army of the Republic of Vietnam forces and the 1st Cavalry and it sustained a terrific pounding from aerial rocket artillery and the air assaults.

As darkness fell, it became evident that the battered enemy was attempting to break contact and withdraw. Its forces rapidly broke down into small groups with some individual soldiers even attempting to get away among the crowds of fleeing refugees. Contacts were made throughout the night with many small enemy groups trying to get out of the city. The enemy had suffered a terrible mauling from the Army of the Republic of Vietnam defenses within Quang Tri and had been thoroughly demoralized by the air assaults, gunships, and aerial rocket artillery of the 1st Cavalry Division. Although well equipped, the North Vietnamese Army troops appeared to be inexperienced and were obviously completely unfamiliar with the airmobile tactics of the 1st Cavalry. The aerial rocket artillery and the helicopter gunships experienced unusual success against the enemy troops. Many of them attempted to play dead as the helicopters approached, seldom attempting to return fire.

By noon on 1 February, Quang Tri City had been cleared of the enemy and the 1st Brigade immediately initiated pursuit. A Company of the 1st Battalion, 502d Airborne made a heavy contact just south of Quang Tri killing 76 of the enemy with the help of aerial rocket artillery. Other units of the 1st Brigade made numerous smaller contacts throughout the day as the brigade elements moved out in ever-increasing concentric circles around the city.

In this abortive attack, the enemy lost 914 soldiers killed in action and 86 captured. The city of Quang Tri was without a doubt one of the major objectives of the *Tet* offensive. Its successful defense was one of the highlights of this period. The enemy's offensive time table in the I Corps Tactical Zone had been completely disrupted and a major communications hub remained in allied hands. The successful defense of Quang Tri resulted from the tenacious defense of the city itself by Army of the Republic of Vietnam forces; the accurate assessment of the tactical situation by Colonel Rattan and Mr. Brewer, the senior province advisor; and the ability of the 1st Cavalry Division to immediately react to this assessment by a complete change of orientation of its units and launching a series of air assaults on top of the enemy's supporting positions that very afternoon.

The 1st Cavalry at Hue

Volumes have been written about the battle for Hue and the house-to-house fighting that went on until almost the end of February. My Cavalry squadron, the 1st Squadron, 9th Cavalry, had been very actively engaged in the outskirts of Hue and the division was given the mission to interdict the routes of egress and destroy the enemy units west of the city. The 2d Battalion, 12th Cavalry began to seal off the city from the west and the north with its right flank on the Perfume River on 2 February. The weather was miserable at this time with ceilings being at most 150 to 200 feet. Nevertheless, helicopters kept flying and placed the troops close to the assault positions even if they could not make an actual air assault. I think it was at this time that General Creighton W. Abrams said that any previous doubts that he had had about the ability of the helicopter to fly in marginal weather were removed.

The 1st Cavalry was spread particularly thin at this time. The 1st Brigade with four battalions was completely occupied at Quang Tri. The base at Camp Evans with approximately 200 helicopters had to be secured and the main land supply line from Dong Ha down to Camp Evans needed to be reinforced.

The logisticians had more than their share of problems during the Battle of Hue. The road from Hue-Phu Bai to Da Nang was cut and we actually backtracked some supplies from the north at Cua Viet. The Air Force did a tremendous job in flying parachute resupply missions to Camp Evans. At times they were dropping supplies with the ceiling around 300 feet using our pathfinders and Ground Control Approach radar. It was eerie to see the parachutes come floating out of the clouds minutes after the C–130's had passed. During this same period, our flying cranes and Chinooks flew out to sea and landed on the AKA's, picked up supplies, and flew them back to Camp Evans. To the best of my knowledge, this was the first example of ship-to-shore resupply in combat.

Two Cavalry battalions were initially committed to the mission at Hue and eventually four battalions were involved in some of the most furious combat that had taken place in Vietnam since the beginning of the war. Air strikes were very difficult to call in because of the bad weather and low ceilings. Most of our helicopter operations were at an altitude of about 25 feet. The Cavalry had cut off one of the enemy's main supply lines and had taken a heavily fortified tactical headquarters at La Chu on the outskirts of the city of Hue. Our Naval gunfire support during the battle of Hue could only be described as superb. For example, one cruiser set a record for the number of shells fired in one day in support of ground operations.

I had sent one of my assistant division commanders, Brigadier General Oscar E. Davis, into the city of Hue to be my personal representative at the headquarters of Brigadier General Ngo Quang Truong, Commanding General of the 1st Army of the Republic of Vietnam Division. General Truong had forecast that "when the Cav reaches the walls of Hue, the battle would be over." He was right. Later, interrogation of prisoners indicated that three enemy regiments had begun moving from around Khe Sanh into the area of Hue between 11 and 20 February to reinforce the weakening local forces. The aggressive actions by the 3d Brigade of the 1st Cavalry, commanded by Colonel Campbell, around La Chu had seriously disrupted the enemy plans for reinforcement.

Summary of Tet

The *Tet* offensive had hurt the enemy severely. The North Vietnamese Army and Viet Cong had lost some 32,000 men killed and 5,800 detained from the period between 29 January and 11 February 1968. They had lost over 7,000 individual weapons and almost 1,300 crew-served weapons.

The communists had paid another kind of price. By choosing *Tet* for their attack, they had alienated a major portion of the population who considered that a sacred time of the year. They had also brought the battle into the very midst of the heavily populated areas causing many civilian casualties who were caught in the cross fire. Most importantly, they had totally misjudged the mood of the South Vietnamese. Believing in their own propaganda, the Viet Cong had called for and expected a popular uprising to welcome the Communists as liberators. Nothing approaching that myth occurred anywhere in Vietnam. The Government of Vietnam did not collapse under the *Tet* offensive. On the contrary, it rallied in the face of the threat with a unity and purpose greater than that which had ever been displayed up to that time.

CHAPTER IX

Major Operations, 1968

Khe Sanh

Weeks before the *Tet* offensive, the eyes of the world had been focused on Khe Sanh as all signs pointed to a major enemy attack on this Marine outpost.

Located some fifteen miles south of the Demilitarized Zone and barely seven miles from the eastern frontier of Laos, the Khe Sanh base functioned primarily as a support facility for surveillance units watching the demilitarized zone and probing the outer reaches of the Ho Chi Minh Trail in nearby Laos. Khe Sanh was almost completely surrounded by towering ridges and stood in the center of four valley corridors leading through the mountains to the north and northwest of the base. To the south Khe Sanh overlooked Highway Nine, the only east-west road in the northern province to join Laos and the coastal regions. The base itself was laid out on a flat laterite plateau. It was shaped somewhat like an irregular rectangle and covered an area approximately one mile long and one-half mile wide. A key feature of the base was a 3,900 foot aluminum mat runway which during favorable weather conditions could accommodate fixed-wing aircraft up to C–130 transports.

The enemy's primary objective in the *Tet* offensive in early 1968 was to seize power in South Vietnam and cause the defection of major elements of its armed forces. In conjunction with this, the enemy apparently expected to seize by military action large portions of the northern two provinces and to set up a "Liberation Government." Khe Sanh was a part of this plan and was obviously an initial objective of the North Vietnamese Army. Its seizure would have created a serious threat to our forces in the northern area and cleared the way for the enemy's advance to Quang Tri City and the heavily populated region. In addition, as General Westmoreland stated, "There is also little doubt that the enemy hoped at Khe Sanh to obtain a climacteric victory such as he had done in 1954 at Dien Bien Phu in the expectation that this would produce a psychological shock and erode American morale."

On the 25th of January I was directed to prepare a contingency plan for the relief or reinforcement of the Khe Sanh Base. This action was the first in a chain of events that was later to emerge as Operation PEGASUS. The mission was three-fold: One, to relieve the Khe Sanh Combat Base; two, to open Highway Nine from Ca Lu to Khe Sanh; and, three, to destroy the enemy forces within the area of operations. After the inevitable delays caused by the *Tet* offensive, the 1st Cavalry Division began preparation for one of its most classical engagements.[1]

In the first weeks of 1968 signs of an impending enemy attack at Khe Sanh continued to mount. As many as four North Vietnamese divisions were identified just north of the Demilitarized Zone. Two of these divisions, the 325th C and the 304th, were thought to be concentrated in the northwestern edge of Quang Tri Province with elements already in position in the hills surrounding Khe Sanh. In addition, there were numerous indications that the enemy was moving up many batteries of artillery in the southern half of the Demilitarized Zone as well as in areas close to the Laos border—all well within range of the Khe Sanh Combat Base.

Convinced that a massive enemy blow would soon fall on Khe Sanh, the American command moved swiftly to strengthen its forces in the area. The 5th Marine Regiment was hastily redeployed from the Da Nang area north to the vicinity of Hue and the U. S. Army's 1st Cavalry Division (Airmobile) was displaced to the northern provinces along with two brigades of the 101st Airborne Division. Beginning in mid-January, the combat base at Khe Sanh was consecutively reinforced by the 2d Battalion of the 26th Marines, the 1st Battalion of the 9th Marines, and finally the 37th South Vietnamese Ranger Battalion, bringing the troop level at the base to a little less than 6,000 men.

Concurrent with the buildup of the allied forces in the vicinity of the Demilitarized Zone, B–52 bombers began to systematically pattern bomb suspected enemy locations near Khe Sanh and tactical fighter bombers stepped up attacks in North Vietnam's southern panhandle. East of Khe Sanh, U. S. Army heavy artillery was assembled at the "Rock Pile" and at Camp Carroll to provide long range fire support to the Khe Sanh base on a quick reaction basis.

In the early morning hours of 21 January the enemy had made his long-awaited move against Khe Sanh. The main base was hit

---

[1] The 1st Cavalry's operation in I Corps Tactical Zone—including the move north, the *Tet* Offensive, securing base areas, and preparation for PEGASUS—had been given the name of Operation JEB STUART.

by withering artillery, rocket and mortar fire and probing efforts against outlying defensive positions to the north and northwest. South of the base the enemy attempted to overrun the villages of Khe Sanh and Huong Hoa, but were beaten back by Marine and South Vietnamese defenders. In this initial action, enemy fire destroyed virtually all of the base ammunition stock as well as a substantial portion of the fuel supplies. In addition, the all-important air strip was severely damaged forcing a temporary suspension of flights into the area

From these beginnings, the battle lines at Khe Sanh were tightly drawn around the main base and its adjacent mountain strongholds. For the next 66 days world-wide attention would remain riveted on Khe Sanh where the enemy seemed to be challenging the United States to a set battle on a scale not attempted since the great communist victory at Dien Bien Phu.

As the siege of Khe Sanh progressed, air delivered fire support reached unprecedented levels. A daily average of 45 B–52 sorties and 300 tactical air sorties by Air Force and Marine aircraft were flown against targets in the vicinity of the base. The U. S. Navy provided additional aircraft sorties from carriers. Eighteen hundred tons of ordnance a day were dumped into the area laying waste to huge swaths of jungle terrain and causing hundreds of secondary explosions. In seventy days of air operations 96,000 tons of bombs were dropped, nearly twice as much as was delivered by the Army Air Force in the Pacific during 1942 and 1943. B–52 Arc Light strikes were particularly effective against enemy personnel and had a great psychological impact on their troops.

Artillery fire provided an important supplement to the air campaign. The sixteen 175-mm guns at Camp Carroll and the "Rock Pile" as well as the forty odd artillery pieces positioned inside the Khe Sanh perimeter directed some 118,000 rounds at enemy positions within a ten mile radius of the base.

But even though the allies were successful in keeping Khe Sanh supplied by air and surrounding its defenders with a pulverizing wall of firepower, a deep feeling of apprehension over the fate of this outpost persisted in official and public circles. After all, the base was completely encircled by an enemy with at least a three to one numerical advantage, and that enemy unmistakably commanded the initiative. Much of the high ground overlooking Khe Sanh was undefended and presumably under the control of the North Vietnamese. Despite intensive counter fires the enemy managed to regularly pound the life-supporting runway and other critical installations with mixed barrages of artillery, rocket, and mortar fire, averaging 150 rounds a day. On 23 February, the enemy fired over 1,300 rounds at the base. Also, weather worked against the Khe Sanh defenders. The base was repeatedly blanketed by fog and low-hanging clouds which not only interrupted fighter air sorties but also permitted the enemy to move men and equipment undetected to the very edges of the Marine defense perimeters.

On 25 January 1968, General Westmoreland had directed that a Military Assistance Command, Vietnam, forward command post be established in the I Corps area. General Abrams would be his representative in this forward Command Post. Military Assistance Command, Vietnam, Forward was activated, staffed, and became operational during the period of the *Tet* Offensive. The staff of Military Assistance Command, Vietnam, Forward would soon form the nucleus of Provisional Corps Vietnam when General Abrams closed his forward command post. General Abrams arrived at the command post in Phu Bai on 13 February and on that same day the decision was made to move the 101st Airborne Division (–) into Military Assistance Command, Vietnam, Forward's area of operation. The 3d Brigade of the 82d Airborne Division, being airlifted from the United States, would come under the 101st Airborne Division upon its arrival in the tactical zone.

Priority of movement for units being moved into the northern I Corps Tactical Zone were established by Military Assistance Command, Vietnam, Forward on 1 March. The 2d Brigade, 1st Cavalry Division was to join the remainder of the division operating in the area north of Hue and south of Quang Tri. Next, the command and control and support elements of the 101st Airborne Division were to move north, and finally, the 3d Brigade, 82d Airborne Division would arrive in country, through Chu Lai, and move north.

The 1st Cavalry Division began repositioning its forces to allow for the arrival of its 2d Brigade and the 101st Airborne Division (–). The 1st Brigade, 1st Cavalry was to continue security of Quang Tri and to conduct operations south and west of the city. The 2d Brigade, 1st Cavalry Division, was to assume responsibility for the operational area of the 2d Brigade, 101st Airborne Division, upon arrival in I Corps Tactical Zone and conduct operations to clear the enemy elements in the Hai Lang–My Chanh area. The 3d Brigade, 1st Cavalry Division, was to continue to operate west of Hue until relieved by the 2d Brigade, 101st Airborne Division. When the 2d Brigade, 101st began operations west and northwest of the City of Hue, the 3d Brigade, 1st Cavalry Division would move north to Camp Evans, headquarters of the division, and assume the mission of base defense.

The roles and missions of Provisional Corps Vietnam were published by Military Assistance Command, Vietnam, in a Letter of Instruction dated 3 March 1968. Lieutenant General William B. Rosson was designated Commanding General of the Corps which, upon its activation on 10 March, was under operational control of III Marine Amphibious Force. The Provincial Corps, Vietnam, was authorized direct coordination with Army of the Republic of Vietnam forces within its area of responsibility.

Operation Pegasus

This, then, is how the situation stood in early 1968. Press correspondents began to dramatize the developments. Repeatedly the public was told that Khe Sanh was likely to be a "very rough business with heartbreaking American casualties." The impending battle was seen as a major test of strength between the U. S. and North Vietnam, with heavy political and psychological overtones.

On 2 March, I went to Da Nang to present our plan for the relief of Khe Sanh to General Cushman, Commanding General, III Marine Amphibious Force. In attendance at this briefing was General Abrams, Deputy Commander, U. S. Military Assistance Command, Vietnam, who still had his advance headquarters at Hue-Phu Bai. Our plan was approved in concept and provisional troop allocations were made.

To accomplish the mission, the 1st Cavalry Division would be augmented by the following non-divisional units: 1st Marine Regiment, 26th Marine Regiment, III Army of the Republic of Vietnam Airborne Task Force, and the 37th Army of the Republic of Vietnam Ranger Battalion. In all, I would have over 30,000 troops under my direct operational control.

Having been given the broad mission and the forces necessary, I was given complete freedom on how to do the job from the beginning. Seldom is a commander so blessed. In the early stages of planning, verbal orders were the modus operandi. As the concept took shape, I asked for representatives from all the units that would be working with us and detailed plans were developed under the critical supervision of Colonel Putnam. We even constructed a sand table model of the Khe Sanh area. I made several trips into the surrounded Marine garrison to coordinate directly with its commander, Colonel David E. Lownds.[2]

Many different elements were involved and all would have to be pulled together under my command on D-day. If we wanted surprise, speed and flexibility during the actual attack, everyone had to understand their part of the plan and the control procedures. This was especially true of firepower. A lot of things were going to be moving through the same air space—bombs, rockets, artillery shells, helicopters and airplanes. We had to assure ourselves that none got together inadvertently.

The basic concept of Operation PEGASUS was as follows: The 1st Marine Regiment with two battalions would launch a ground attack west toward Khe Sanh while the 3d Brigade would lead the 1st Cavalry air assault. On D+1 and D+2 all elements would continue to attack west toward Khe Sanh; and, on the following day, the 2d Brigade of the Cavalry would land three battalions southeast of Khe Sanh and attack northwest. The 26th Marine Regiment, which was holding Khe Sanh, would attack south to secure Hill 471. On D+4, the 1st Brigade would air assault just south of Khe Sanh and attack north. The following day the 3d Army of the Republic of Vietnam Airborne Task Force would air assault southwest of Khe Sanh and attack toward Lang Vei Special Forces Camp. Linkup was planned at the end of seven days.

It became evident during the planning that the construction of an airstrip in the vicinity of Ca Lu would be a key factor for the entire operation. This airstrip, which became known as landing zone STUD, had to be ready well before D-day (1 April 1968). Also, it was necessary to upgrade Highway Nine between the "Rock Pile" and Ca Lu to allow prestocking of supplies at landing zone STUD.

I sent one of my assistant division commanders, Brigadier General Oscar E. Davis, to personally supervise the establishing of landing zone STUD as our advance base for PEGASUS. Calling this a "landing zone" is a gross understatement, for landing zone STUD would have to be a major air terminal, communications center, and supply depot for the future.

The 1st Cavalry Division engineers, the Seabees USN Mobile Construction Battalion #5, and the 11th Engineer Battalion,

---

[2] In order to correct an impression given by the newsreel coverage at this time, I must point out that the only "safe" way to get into the Khe Sanh Combat Base was by helicopter. I usually chose to land in the Special Forces area. The C–130's were either delivering their loads by low altitude extraction or by parachute. The runway was the most dangerous and exposed area at Khe Sanh.

A Blue Team Rifle Squad From the 1st Squadron, 9th Cavalry Exiting From a Huey Helicopter

USMC, initiated construction of the C–7A airfield and parking ramp, logistical facilities, and a bunker complex at landing zone Stud on 14 March. By D–6 they had finished an airstrip 1500 feet long by 600 feet wide, ammunition storage areas, aircraft and vehicle refueling facilities, and extensive road nets into the vicinity of landing zone Stud. The Seabees, which had been augmented with very heavy equipment, accomplished the lion's share of the work on the airfield.

Having established a forward base of operations, the second key element to the success of this plan was the closely integrated reconnaissance and fire support effort of the 1st Squadron, 9th Cavalry, under the brilliant leadership of Lieutenant Colonel Richard W. Diller, and air, artillery, and B–52 Arc Light strikes, during the period D–6 to D-day. This was almost a flawless demonstration of properly preparing a battlefield when tactical intelligence was not available.

This is not to say there was not a tremendous intelligence effort focused around Khe Sanh Combat Base itself. In addition to the aerial observation and daily photographic coverage, General Westmoreland had personally made the decision to divert new acoustic sensors from their intended implacement along the DMZ to the approaches around Khe Sanh. Through a complex computer system, these devices could provide early warning of any intrusion and were often used to target B–52 strikes. However, the acoustical sensor system, which was focused on the immediate area of Khe Sanh, did not directly help develop the complete intelligence picture necessary for our proposed attack along Highway Nine.

The actual intelligence on the enemy in the area was very vague and expressed in generalities. The 1st Squadron, 9th Cavalry operated from landing zone STUD in gradually increasing concentric circles up to the Khe Sanh area, working all the time with air cover from the 7th Air Force or the 1st Marine Air Wing. The Cavalry Squadron was almost the only means available to pinpoint enemy locations, antiaircraft positions, and strong points that the division would try to avoid in the initial assaults. The squadron was also responsible for the selection of critical landing zones. Their information proved to be timely and accurate.

During the initial surveillance efforts it became evident that the enemy had established positions designed to delay or stop any attempt to reinforce or relieve Khe Sanh. Positions were identified on key terrain features both north and south of Highway Nine. As part of the reconnaissance by fire, known or suspected enemy antiaircraft positions and troop concentrations were sought out and destroyed either by organic fire or tactical air. Landing zones were selected and preparations of the landing zones for future use were accomplished by tactical air using specially fused bombs and B–52 Arc Light strikes. During this phase of the operation, the 1st Squadron, 9th Cavalry developed targets for 632 sorties of tactical air, 49 sorties for the specially fused bombs, and twelve B–52 Arc Light strikes. The thoroughness of the battlefield preparation was demonstrated during the initial assaults of the 1st Cavalry Division, for no aircraft were lost due to antiaircraft fire or enemy artillery.

At this point, I must mention the element of surprise. Certainly the enemy knew we were in the area. Our own reporters let the whole world know the situation as they saw it, and the arm-chair strategist could ponder the problem each evening in front of his color TV. However, the inherent capabilities of the airmobile division presented the enemy with a bewildering number of possible thrusts that he would have to counter, all the way to the Laotian

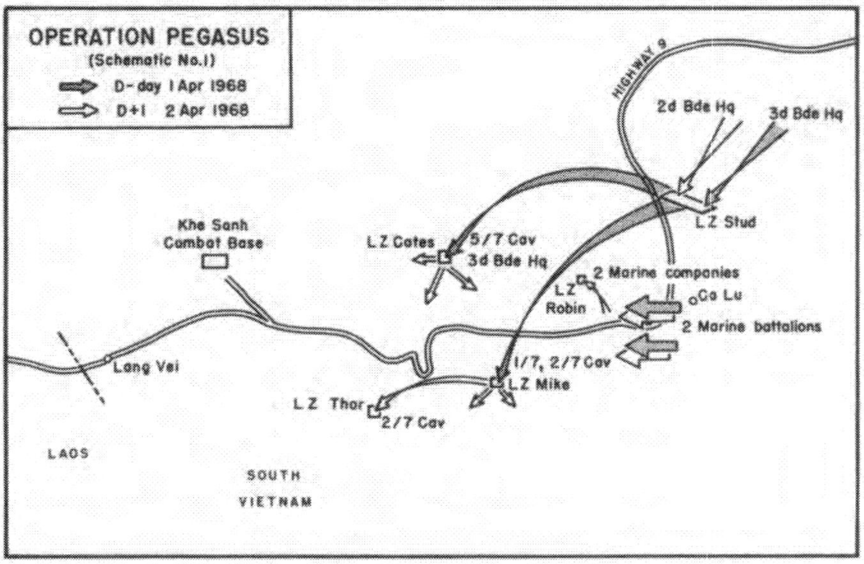

MAP 4

border. Also, there would be a major diversionary attack in the vicinity of the DMZ on D–1. The initiative was ours.

At 0700 on 1 April 1968 the attack phase of Operation PEGASUS commenced as two battalions of the 1st Marine Regiment under Colonel Stanley S. Hughes attacked west from Ca Lu along Highway Nine. (*Map 4*) The 11th Marine Engineers followed right on their heels. At the same time, the 3d Brigade of the 1st Cavalry under Colonel Hubert (Bill) S. Campbell was airlifted by Chinooks and Hueys into landing zone STUD in preparation for an air assault into two objective areas further west. Weather delayed the attack until 1300, when the 1st Battalion, 7th Cavalry, commanded by Lieutenant Colonel Joseph E. Wasiak, air assaulted into landing zone MIKE located on prominent ground south of Highway Nine and well forward of the Marine attack. Lieutenant Colonel Roscoe Robinson, Jr., led the 2nd Battalion, 7th Cavalry into the same landing zone to expand and develop the position. The 5th Battalion, 7th Cavalry, commanded by Lieutenant Colonel James B. Vaught, air assaulted into an area north of Highway Nine approximately opposite landing zone MIKE.

These two objectives had been chosen after careful reconnaissance by the Cavalry Squadron indicated no major enemy defenses. Though almost halfway to Khe Sanh, they were within range of supporting artillery. Both landing zones were secured and no significant enemy resistance was encountered. A battery of 105-mm howitzers was airlifted into each landing zone and Colonel Campbell moved his brigade headquarters into the northern landing zone, landing zone CATES. Bad weather notwithstanding, everything was in place prior to darkness. The major accomplishment of D-day was the professional manner in which this tremendously complex operation, with all its split-second timing and coordination, had to be delayed several hours yet was completed as planned.

The bad weather of D-day was to haunt the 1st Cavalry throughout Operation PEGASUS. Seldom were airmobile moves feasible much before 1300. "Good weather" was considered to be any condition when the ceiling was above 500 feet and slant range visibility was more than a mile and a half. The bad weather further proved the soundness of establishing landing zone STUD as the springboard for the assaults. Troops, ammo and supplies could be assembled there ready to go whenever the weather to the west opened up. Marshalling areas further away would have drastically deteriorated response time.

On D+1 (2 April), the 1st Marine Regiment continued its ground attack along the axis of Highway Nine. Two Marine companies made limited air assaults to support the Regiment's momentum. The 3d Brigade air assaulted the 2d Battalion, 7th Cavalry into a new position further to the west while the other two battalions improved their positions. The 2d Brigade under Colonel Joseph C. McDonough moved into marshalling areas in preparation for air assaults the next day, if called upon.

Lest all of this sound routine, I want to emphasize that only the initial assaults on D-day were fixed in time or place. All subsequent attacks were varied to meet changes in the enemy situation or to capitalize on unexpected progress. As an example, I ordered an acceleration of the tempo when the results of D-day attacks gave clear evidence that the enemy was unprepared.

Our initial thrusts had met less enemy resistance than expected. As a consequence, the 2d Brigade was thrown into the attack a day earlier than the original schedule with three battalions [3] moving into two new areas south and west of our earlier landing zones.

---

[3] The 1st Battalion, 5th Cavalry under Lieutenant Colonel Robert L. Runkle (who was killed in action the following day and replaced by Lieutenant Colonel Clarence E. Jordan), the 2d Battalion, 5th Cavalry under Lieutenant Colonel Arthur J. Leary, Jr., and the 2d Battalion, 12th Cavalry under Lieutenant Colonel Richard S. Sweet.

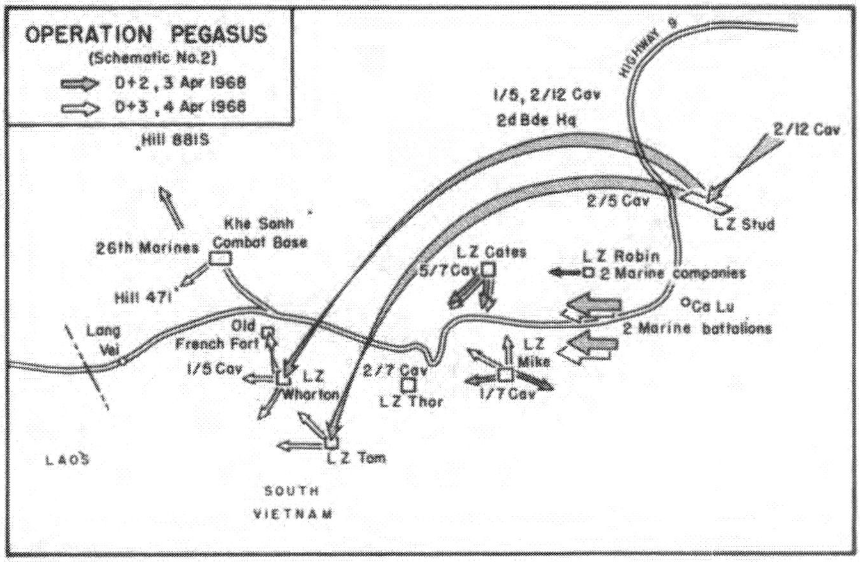

MAP 5

(*Map 5*) They received enemy artillery during the assaults, but secured their objectives without serious difficulty. We now had six air cavalry battalions and supporting artillery deep in enemy territory.

I was anxious to get the 26th Marine Regiment out of their static defense position as soon as feasible; so, on D+3, I ordered Colonel Lownds to make a battalion-size attack south from Khe Sanh to seize Hill 471, a strategic piece of terrain affording a commanding view of the base. Following a heavy artillery preparation, the Marines successfully seized the hill killing thirty of the enemy. On the same day, the 2d Brigade of the Cavalry Division assaulted one battalion into an old French fort south of Khe Sanh. Initial contact resulted in four enemy killed. The remaining uncommitted brigade was moved into marshalling areas.

On D+4 (5 April), the 2d Brigade continued its attack on the old French fort meeting heavy enemy resistance. Enemy troops attacked the Marines on Hill 471 but were gallantly repulsed with 122 of the enemy left dead on the battlefield. The tempo of this battle was one of the heaviest during the operation. The 3d Airborne Task Force, Army of the Republic of Vietnam, was alerted to prepare to airlift one rifle company from Quang Tri to effect linkup with the 37th Army of the Republic of Vietnam Ranger Battalion located at Khe Sanh. Units of the 1st Brigade under

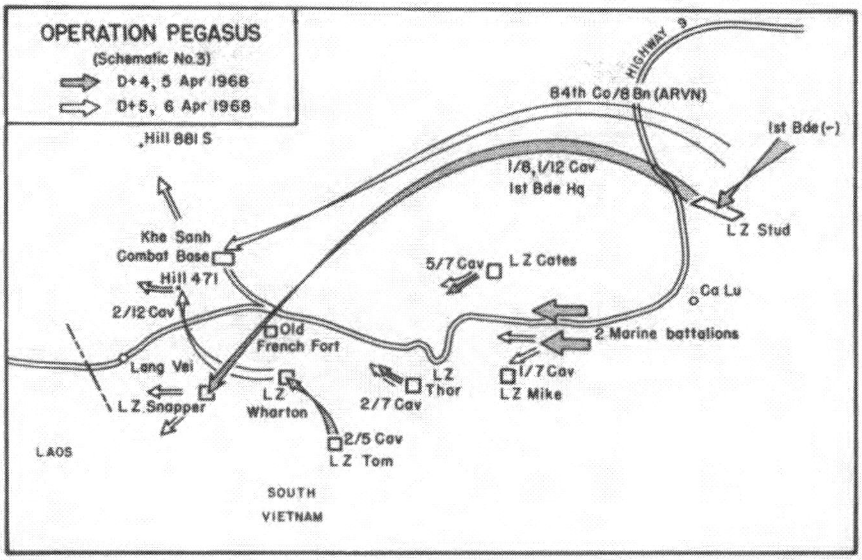

MAP 6

Colonel Stannard entered the operation with the 1st Battalion, 8th Cavalry, commanded by Lieutenant Colonel Christian F. Dubia, air assaulting into landing zone SNAPPER, due south of Khe Sanh and overlooking Highway Nine. The circle began to close around the enemy. (*Map 6*)

On D+5 (6 April), the 1st Marine Regiment continued its operations on the high ground north and south of Highway Nine, moving to the west toward Khe Sanh. The heaviest contact on that date occurred in the 3d Brigade's area of operation as the 2d Battalion, 7th Cavalry under the inspired leadership of Colonel Robinson continued its drive west on Highway Nine. Enemy blocking along the highway offered stubborn resistance. In a day-long battle which ended when the enemy summarily abandoned his position and fled, the battalion had accounted for 83 enemy killed, one prisoner of war captured, and 121 individual and ten crew-served weapons captured. The troops of the 1st Cavalry Division were airlifted to Hill 471 relieving the Marines at this position. This was the first relief of the defenders of Khe Sanh. Two companies of troopers remained on the hill while two other companies initiated an attack to the south toward the Khe Sanh Hamlet.

We had plotted heavy enemy artillery that had been dug deeply into the rocks of the Co Roc Mountains in Laos just west of Lang Vei. As we neared Khe Sanh I was concerned that these 152-mm guns could bring our landing zones under fire at any time. But, we were forbidden to cross the border and the heaviest aerial bombs could not dislodge these positions. They remained a threat throughout PEGASUS.

The 1st Cavalry forces on landing zone SNAPPER were attacked by an enemy force utilizing mortars, hand grenades, and rocket launchers. The attack was a disaster for the enemy and twenty were killed. At 1320 the 84th Company of the Vietnamese 8th Airborne Battalion was airlifted by 1st Cavalry Division aircraft into the Khe Sanh Combat Base and linked up with elements of the 37th Ranger Battalion. The lift was conducted without incident and was marked as the official link-up in forces at Khe Sanh.

On 7 April the South Vietnamese III Airborne Task Force air assaulted three battalions into positions north of the road and east of Khe Sanh to block escape routes toward the Laotian border. Fighting throughout the area was sporadic as the enemy attempted to withdraw. American and South Vietnamese units began picking up significant quantities of abandoned weapons and equipment. The old French fort which was the last known enemy strong point around Khe Sanh was completely secured.

At 0800 on 8 April the relief of Khe Sanh was effected and the 1st Cavalry Division became the new landlord. The 3d Brigade airlifted its command post into Khe Sanh and Colonel Campbell assumed the mission of securing the area. This was accomplished after the 2d Battalion, 7th Cavalry successfully cleared Highway Nine to the base and effected link-up with the 26th Marine Regiment. The 3d Brigade elements occupied high ground to the east and northeast of the base with no enemy contact. At this time it became increasingly evident, through lack of contact and the large amounts of new equipment being found indiscriminately abandoned on the battlefield, that the enemy had fled the area rather than face certain defeat. He was totally confused by the swift, bold, many-pronged attacks. Operations continued to the west.

On 9 April all 1st Marine Regiment objectives had been secured and Highway Nine was repaired and secured with only scattered incidents of enemy sniper fires. Enemy mortar, rocket and artillery fire into Khe Sanh became increasingly sporadic.

On the following day the 1st Battalion of the 12th Cavalry, commanded by Lieutenant Colonel Robert C. Kerner, under the 1st Brigade seized the old Lang Vei Special Forces Camp four miles west of Khe Sanh against light enemy resistance. This was the site of an enemy attack in mid-February when North Vietnamese troops, supported by armor, overran the camp. Early on the 10th a helicopter from A Troop, 1st Squadron, 9th Cavalry, had located a PT–76 tank and had called in a tactical air strike on the vehicle. The tank was destroyed along with fifteen enemy troops.

Though this was the only recorded tank kill during Operation PEGASUS, we had had intelligence of enemy armor throughout this area. The enemy's offensive at Lang Vei Special Forces Camp had given undeniable proof of this enemy capability and, since that time, intelligence sources estimated a possible company-size armor unit near Khe Sanh. The 1st Squadron, 9th Cavalry had sighted tank treads several times in their early reconnaissance before D-day.

Before the operation I had directed the division to be prepared to use the SS–11 missile system during PEGASUS. This system, which employed a wire-guided armor-piercing missile had been in the theater since the arrival of the 1st Cavalry. However, the lack of lucrative targets had reduced its usefulness. The system had been standardized in the U. S. Army since 1960 when it replaced the lighter French SS–10 missile. Since then hundreds of gunners had been trained at Fort Rucker, Alabama, in the use of the SS–11. We were never able to target this system during PEGASUS, but I still felt it was a very valuable capability that should be maintained in constant readiness.

Highway Nine into the Khe Sanh Combat Base was officially opened on 11 April after the Marine engineers had worked day and night to complete their task. In eleven days the engineers had reconstructed over fourteen kilometers of road, repaired or replaced nine bridges, and constructed seventeen bypasses. Numerous sections of the road had to be cleared of landslides and craters.

I had scheduled more than 38 additional operations to extend our control of the Khe Sanh area but, without warning, on the morning of 10 April I received orders from General Rosson to make plans to extract the Division as soon as possible to prepare for an assault into the A Shau Valley. Advance units started pulling out on the 11th. Limited operations continued until 15 April when Operation PEGASUS was officially terminated.

There was great potential for the continued air assault operations that were abruptly brought to close. The enemy was vulnerable; he was abandoning his equipment; and, he was completely disorganized. The decision to expedite our withdrawal immediately upon completing our primary mission–the relief of Khe Sanh Combat Base–was predicated on a long-range forecast which predicted April as the last possible time for air assault operations in the A Shau Valley before the heavy monsoon rains.

Operation PEGASUS–LAM SON 207A from its inception to its final extraction from the area of operations will long stand as a classic example of airmobile operations. The operation dramatically illustrated the speed and effectiveness with which a large force can be employed in combat using airmobile tactics and techniques. The enemy's repeated failure to quickly comprehend the quick reaction time and capabilities of the 1st Cavalry Division led to his defeat, forced withdrawal, and eventual rout from the battlefield. The enemy was helpless and confused, suffered great losses of men and equipment, and failed in his mission to block and delay the relief of Khe Sanh.

No summation of Operation PEGASUS would be complete without mention of the great team effort of all the Services–Army, Navy, Marines and Air Force. The operation was an ideal example of the synchronization of massive B–52 strikes, tactical air support and artillery firepower with ground maneuver. The South Vietnamese troops gave a splendid performance. The fact that we were able to co-ordinate all of these operations in a single headquarters was a commander's dream. There was no question of command or who was calling the signals. Equally important, I had the full support of General Rosson, who commanded the Provisional Corps Vietnam, and of General Cushman, Commanding General, III Marine Amphibious Force.

The success of the PEGASUS operation can largely be attributed to the detailed planning and preparation that occurred prior to D-day and the effective reconnaissance and surveillance of the area of operations provided by the air cavalry squadron. This reconnaissance and its ability to develop hard targets for the tactical air and B–52 ARC LIGHT strikes cannot be overestimated. The concept of building landing zone STUD as a pivot point for the entire operation proved sound. This base provided a continuous flow of needed supplies and equipment to forward elements of the division. The success of the initial battalion air assaults was rapidly exploited by aggressive company and even platoon-size air assaults, all supported by artillery and air. The enemy, although well dug in, well supplied, and with an initial determination to deter the relief of Khe Sanh, found himself surrounded with no choice but to retreat in rout order back into Laos, leaving behind 1,304 dead and much valuable equipment strewn over the battlefield.

The total success of the operation can be best measured by the mission accomplished. For the first time, the Cavalry had made an air assault as a division entity; every committed battalion came into combat by helicopter. In fifteen days, the division had entered the area of operations, defeated the enemy, relieved Khe Sanh, and been extracted from the assault—only to assault again four days later into the heart of the North Vietnamese Army's bastion in the A Shau Valley.

9th Division in the Delta

While the 1st Cavalry Division was occupied in preparation for PEGASUS and in the actual relief of Khe Sanh, the U. S. 9th Infantry Division was joined with the enemy in an entirely different type of operation in the Mekong Delta. On 1 March 1968, in conjunction with several Army of the Republic of Vietnam units, elements of the 9th Infantry Division began Operation TRUONG CONG DINH in Dinh Tuong and Kien Tuong Provinces in the IV Corps Tactical Zone. By 19 April, they had accounted for 1,716 enemy killed and 999 detained, while the U. S. forces had only lost 57 men. The division, which was operating on water almost as much as on land, saw airmobility in an entirely different light than those troops fighting in the mountains and jungles in the north.

The Delta region in the IV Corps Tactical Zone stretches from the Cambodian border to the tip of the Ca Mau Peninsula. It is a heavily silted level plain area with elevations not in excess of 9 feet above sea level except in the far western area. The entire area is subject to frequent flooding. Extensive embankments had been built over the centuries to channel water into the fertile rice-producing fields. Mud flats and mangrove swamps encircled the Delta regions along the ragged coastline. Road networks were limited but hard-surfaced major roads did exist. Most of the canals carried a heavy burden of the traffic throughout the area.

The 9th Division's operations in the Delta proved to be a unique testing ground for certain of the airmobile concepts. To begin with, the Division found itself on the low order of priority for airmobile assets and, consequently, could compare its operations without airmobility to those with airmobility.

Major General Julian J. Ewell was to recall later:

> To begin with, the 9th Division worked under two rather unusual conditions which led it into a somewhat different approach to airmobile operations than other units. The Division ended up 1968 in the northern Delta area, south of Saigon, where the terrain was quite low, water-soaked and very open. As a result, ground-pounding operations were extremely difficult to conduct, and the open terrain gave one considerable flexibility in conducting airmobile operations. In addition, the enemy (although able to move freely at night) was more or less pinned

down during the day, thus leading to a situation where if he could be found, he could be dealt with fairly readily. Another factor that affected our approach was that the 9th was rather off at the end of the line, and its location, plus its low priority, meant that it usually received a rather austere level of aircraft resources. These two factors combined led to a situation where you had to squeeze maximum results out of a relatively small number of aircraft. However, in late '68, the priority of Delta operations was raised, and the Division began to receive three lift companies and three air cav troops (including its own). We, therefore, found that we had a reasonable number of aircraft in a highly sympathetic environment and were used to getting the most out of our aircraft. This is all background to assure you that we don't make claims to "inventing the wheel" but were in a situation where we had to do what we did to stay alive.

The 9th Infantry Division made a study of its operations from March through August 1968 to analyze division operations with and without airmobile assets. They attempted to quantify the division's effectiveness by stripping out all the other variables with the exception of the addition of helicopter lift and the air cavalry. The study considered that the simplest and most relevant statistical index of combat effectiveness was the average number of Viet Cong losses inflicted daily by the unit in question. This criteria had to be adjusted to account for the fact that units do not engage in offensive field operations every day. This was especially true for the riverine units in the 2d Brigade, 9th Division, which had to allow for extended periods for boat maintenance. In their study, it was assumed that the provision of all types of airmobile resources except the assault helicopter units and air cavalry units was uniform and thus did not materially influence combat effectiveness from day to day.

A total of 313 brigade-days were analyzed. With no airmobility or air cavalry support the 3d Brigade of the 9th Division averaged 0.21 significant enemy contacts per day spent on field operations. However, when supported by an assault helicopter company and an air cavalry troop, brigade performance more than doubled. In other words, with no air assets a brigade made significant contact with the enemy only once every five days; with airmobile assets, it developed contact every other day.

An analysis of the Viet Cong losses per field day produced more definitive inferences. With no air assets, the brigade performance averaged 1.6 Viet Cong losses per field day—hardly a creditable return. However, when a brigade was supported by an air cavalry troop and a helicopter company, the brigade performance rose to 13.6 Viet Cong per day—an increase of 850 percent. The study went on to refine its perimeters, but the conclusion was inescapable. There was an astonishing improvement in the combat effectiveness of the 9th Infantry Division when it was supported by airmobile and air cavalry assets. The last paragraph of the analysis bears repeating:

> It is possible that not all divisions in Vietnam would generate statistical results similar to those of the 9th Division. The inundated flatlands of the Mekong Delta have a double impact on the spread of effectiveness between foot and airmobile operations. Marshy swamps and flooded rice paddies severely penalize ground troops. Units frequently are able to move no more than 500 meters per hour or less. On the other hand, the broad stretches of virtually flat Delta country provide an ideal environment for the unrestricted employment of Army aviation. Presumably such would not be the case in other areas of Vietnam. Statistics for other divisions might be expected to show improvements in combat effectiveness with air assets on the order of two or three to one, depending upon the circumstances. From the standpoint of II Field Force Vietnam, until analyses are made for the other divisions, it would appear worthwhile to allocate an additional assault helicopter company and an air cavalry troop to the 9th Infantry Division.

The A Shau Valley

On 10 April 1968 at landing zone STUD, General Rosson, the commander of Provisional Corps Vietnam, told me to plan immediate movement of the 1st Cavalry into the A Shau Valley. Though tentative plans had been made for operations against this enemy redoubt sometime before, I had no warning that the 1st Cavalry Division would be committed so quickly on the heels of our operations at Khe Sanh. The following day, we began extracting troops from Operation PEGASUS back into our base areas at Quang Tri City and Camp Evans.

The A Shau Valley lies between two high mountain ranges on the western edge of the Republic of Vietnam. On both sides of the valley the mountains climb to over 1,000 meters with the angle of slope varying from 20 to 45 degrees. The Laotian border is less than ten kilometers away. Three abandoned airfields were spread along the valley floor which runs northwest to southeast. The North Vietnamese forces had been in control of the valley since March 1966 when they overran the Special Forces camp in the southern end. Since that time they had built a major base for the infiltration of personnel and supplies from North Vietnam through Laos along Route 547 into Thua Thien Province and the northern I Corps Tactical Zone. (*Map 7*)

Final preparations for Operation DELAWARE–LAM SON 216 were conducted during the last days of Operation PEGASUS. The 1st

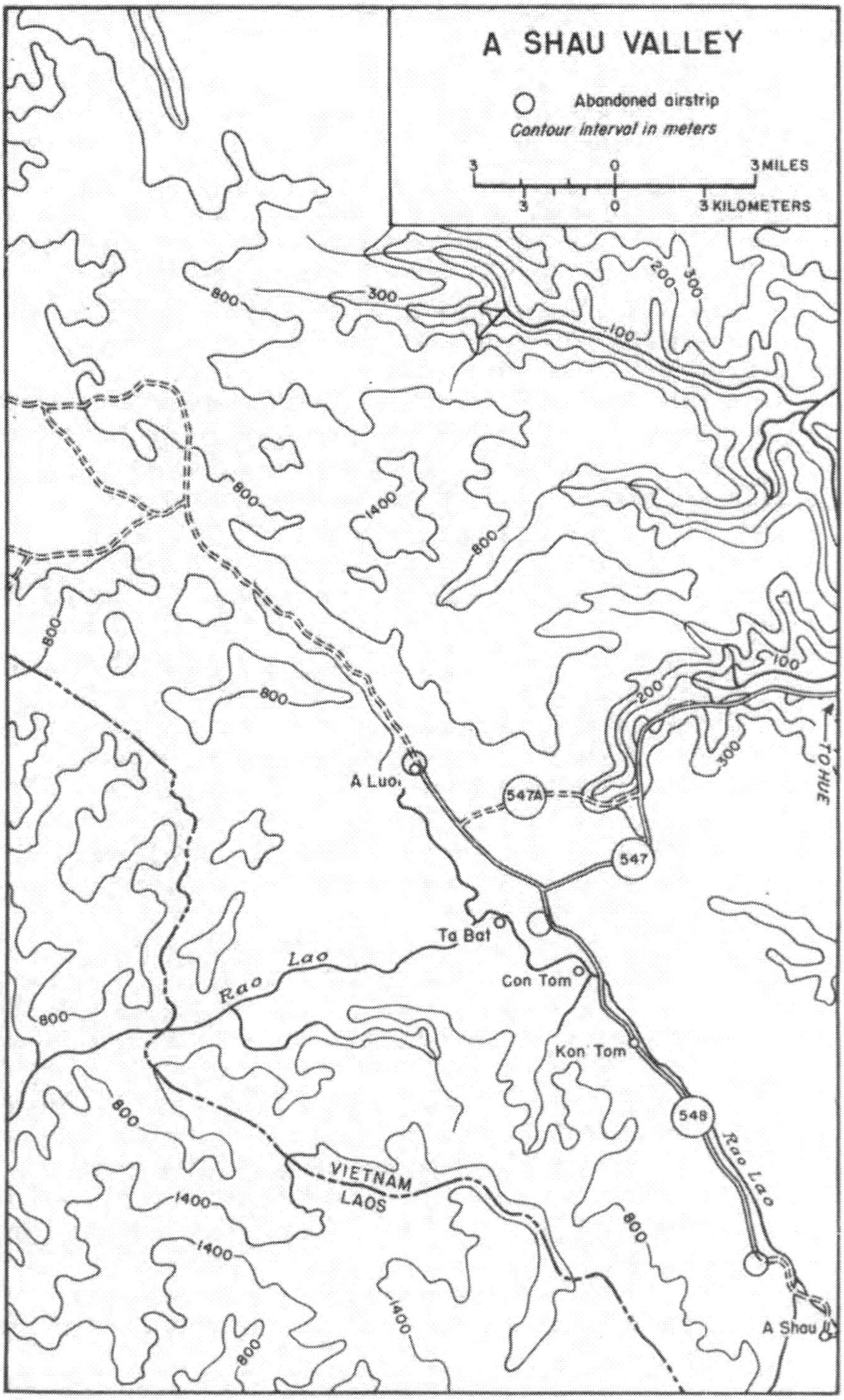
A SHAU VALLEY
Abandoned airstrip
Contour interval in meters
3
0
3 MILES
3
0
3 KILOMETERS
800
300
200
300
100
800
1400
800
800
800
100
200
300
TO HUE
A Luoi
547A
547
Ta Bat
Con Tom
Rao Lao
800
Kon Tom
800
548
Rao Lao
VIETNAM
LAOS
1400
1400
800
1400
800
A Shau

MAP 7

Squadron, 9th Cavalry of the 1st Cavalry Division began an extensive aerial reconnaissance of the A Shau Valley to select flight routes, locate antiaircraft and artillery weapons, and to develop targets for tactical air and B–52 strikes. During the period 14 to 19 April, over 100 B–52 sorties, 200 Air Force and Marine fighter sorties, and numerous aerial rocket artillery missions were flown against targets in the valley. The 1st Brigade of the 101st Airborne Division and the Army of the Republic of Vietnam Airborne Task Force were moved into pre-assault positions ready to make a separate attack on D-day east of the A Shau.

The 101st Airborne Division's role in Operation DELAWARE was to complement the 1st Cavalry Division's assault into the valley itself. The 1st Brigade of the 101st, in coordination with the 3d Army of the Republic of Vietnam Airborne Task Force, was to conduct ground and airmobile assaults to interdict the enemy's routes of withdrawal and infiltration in the area around the junction of Routes 547 and 547A. The 101st Airborne Division's Fire Support Base BASTOGNE, far to the east on Route 547, would just be able to reach the north end of the A Shau Valley with its long range 175-mm guns. The Army of the Republic of Vietnam Airborne Task Force was to be responsible for clearing Route 547A.

Operation DELAWARE was to be a coordinated airmobile and ground attack on two axes using elements of three divisions–the 1st Cavalry, the 101st Airborne, and the 1st Army of the Republic of Vietnam Division. One prong was to be the attack along and astride Routes 547 and 547A, while the main attack was the assault into A Luoi and Ta Bat on the valley floor.

Operation DELAWARE differed from PEGASUS in that during PEGASUS the Cavalry Division had control of all U. S. ground tactical elements. Operation DELAWARE was under the tactical control of Provisional Corps, and my relationship with the 101st Airborne Division was one of coordination. I had also expected to have the Army of the Republic of Vietnam Airborne Task Force as part of our maneuver force, but this unit was shifted to the operational area of the 101st during the latter planning stages. D-day had been tentatively set for 17 April, but I made the condition that it was really contingent on my having three full days of operations in the A Shau Valley by the 1st Squadron, 9th Cavalry in coordination with tactical air and B–52 Arc Light strikes. The purpose of this effort was to determine and neutralize the heavy enemy antiaircraft concentrations.

By the 16th of April, the 1st Squadron, 9th Cavalry still had not had three full days of good weather to operate in the valley, so

I recommended to General Rosson that D-day be postponed until the 19th in which he concurred. In my initial plan I had selected the area around A Luoi for the landing zones in which the 1st Brigade would make the initial assault, the 3d Brigade having made the initial assaults into Operation PEGASUS. The area around A Luoi offered better open landing zones and it gave us immediate control of the airstrip which we could improve for future operations. However, during the final days of the reconnaissance by the Air Cavalry Squadron, I could not get any assurance that an assault into the A Luoi area would not be very costly because they had encountered the heaviest antiaircraft fire right in that area. Though much had been neutralized by air strikes, new positions appeared each day and it became obvious that an alternate plan should be considered. The alternate I selected was an assault into the northern part of the valley by the 3d Brigade, since the 3d Brigade had been prepared to go into this area from the beginning. This alternate plan had several advantages. By going north first, we would immediately cut off the entries into the valley from the new highway coming in from Laos which the enemy had constructed. But, most of all, the threat from antiaircraft fire was less in this area and, after the initial assault, artillery could be placed in the northern landing zones to further neutralize the antiaircraft fire in the center of the valley. The disadvantages of this plan included the fact that we would have to use extremely limited landing zones and that we would delay securing the A Luoi airstrip.

For Operation DELAWARE the 1st Cavalry would have its own 1st and 3d Brigades and coordinate with the 3d Army of the Republic of Vietnam Regiment. The 2d Brigade of the Cavalry remained in the Khe Sanh area to secure the base and continue limited operations in that area. This left the Camp Evans area almost completely void of combat units. Consequently, the 196th Light Infantry Brigade of the Americal Division was given the mission of defense of the base at Camp Evans under operational control of the 1st Cavalry Division. As a further mission they were also designated as a reserve unit for the Corps.

On the morning of the 19th of April the 3d Brigade of the 1st Cavalry Division under Colonel Campbell made the initial assault into the A Shau Valley. Prior to the assault, six B–52 strikes had been delivered in the northern part of the valley and two strikes delivered on the main roads to the east. Tactical Air and artillery hit numerous targets just before the helicopters set down. Despite the large amount of preparatory fire, enemy antiaircraft fire was intense.

The 5th Battalion, 7th Cavalry air assaulted into landing zone TIGER near the winding road into Laos, and was soon followed by its direct support artillery battery. The 1st Battalion, 7th Cavalry went into landing zone VICKI on the slope northeast of TIGER. Although the initial assaults of the 5th Battalion, 7th Cavalry and 1st Battalion, 7th Cavalry were virtually unopposed by ground action, subsequent assault lifts received intense antiaircraft fire. In these later assaults, 23 helicopters were hit by ground-to-air fire and ten aircraft were destroyed. Because of the intense antiaircraft fire, deteriorating weather conditions, and the extensive engineering effort required to prepare artillery positions at landing zone VICKI, insertion of the direct support artillery battery into landing zone VICKI was aborted for that day.

To the east, the 1st Brigade of the 101st Airborne Division initiated operations out of Fire Support Base BASTOGNE with one battalion attacking to the southwest. Later in the morning of the 19th, another battalion air assaulted into the landing zone near the junction of Routes 547 and 547–A. These battalions received light to moderate contact throughout the day.

On D+1, the 3d Brigade of the 1st Cavalry continued to deploy into the northern A Shau Valley and to spread their area of operations. The 1st Battalion, 7th Cavalry under Colonel Wasiak attacked to the southeast out of their original landing zone; the 2d Battalion, 7th Cavalry under Colonel Robinson began an air assault to establish a landing zone further south in the valley; and the 5th Battalion, 7th Cavalry moved to interdict Route 548 which entered the A Shau Valley from Laos.

During the first few days of Operation DELAWARE, in spite of very low ceilings, thunder storms, and heavy enemy antiaircraft fire, the 1st Cavalry Division's helicopters and the U. S. Air Force's C–130 aircraft flew repeated missions into the valley to deliver the required supplies. During the first days of the operation, navigation was strictly by pilotage.

The weather during Operation DELAWARE was almost unbelievably bad. Heavy clouds, fog, thunder storms, and low ceilings made heroic feats of airmanship almost commonplace. Not only were the conditions bad in the area of tactical operations, but even the departure area from Camp Evans had conditions which usually forced our helicopters to climb up through an overcast on instruments, reassemble a formation on top of the clouds, fly to the target area, and then search for some sort of hole in the clouds to make their descent. What should have been a simple twenty-minute flight was usually an hour and twenty minutes of stark terror. Our young warrant officer aviators and our junior officer aviators flew day in and day out in Hueys, Chinooks, and cranes at the very limit of their capabilities. The operation was a phenomenal piece of flying, but from a commander's viewpoint it was sheer agony to see what my people had to go through to accomplish our mission.

The deployed battalions of the 1st Cavalry Division were making sporadic contact with small enemy units on 21 April. Shortly after noon, Company B of the 1st Battalion, 7th Cavalry, discovered an enemy maintenance area containing two trucks, two bulldozers, and assorted engineering equipment. Markings on the bulldozers gave clear evidence that they had been manufactured in the USSR.

With improvements in weather conditions on 22 April, the 3d Brigade, 1st Cavalry, continued to improve their bases in northern A Shau Valley and to bring in needed supplies and equipment. The 1st Battalion of the 7th Cavalry had completed an overland attack and secured a landing zone further south. This battalion was now in position to support the coming assault into the A Luoi Airfield and the central portions of the A Shau Valley.

On 24 April the 2d Battalion, 8th Cavalry, under Lieutenant Colonel John V. Gibney; led the assault of Colonel Stannard's 1st Brigade into a landing zone two kilometers south of the A Luoi Airfield. The 1st Battalion, 12th Cavalry and 1st Battalion, 8th Cavalry completed the Brigade movement the following day when they air assaulted the airfield itself. The 1st Brigade then began reconnaissance in force operations moving to the south and west. Numerous caches of enemy communications equipment, vehicles, ammunition, and weapons were uncovered. Operation DELAWARE was spoiling the enemy's supply depots in the A Shau Valley. The 1st Cavalry troopers were also finding on the ground the heavy antiaircraft weapons that had hit them so hard during the air assaults. Many antiaircraft weapons were uncovered plus thousands of rounds of 23-mm and 37-mm ammunition.

The 1st Brigade of the 1st Cavalry Division continued their buildup at the A Luoi airfield by flying in heavy engineer equipment, sectionalized in small enough loads to be lifted by crane helicopters. The cranes were fueled with just sufficient JP–4 to make the round trip in order to have sufficient lift capability to sling load this heavy equipment over the ridge lines.

On 26 April the buildup at A Luoi continued with fifteen air drops of resupply from C–130 aircraft. The 1st Cavalry Division

Forward Command Post was established near the airfield and the 1st Brigade's forward support element was air lifted in. The forward support element and an Air Force tactical air control party worked together in a joint operation continuing daily resupply by heavy drop. This reduced the demand on the division's aviation assets to complete unit distribution within the valley. However, a large amount of the resupply for the 3d Brigade elements continued to be lifted from Camp Evans by organic aircraft. At 1400 on 26 April a C-130 received enemy antiaircraft fire after breaking out of the overcast too far to the south of the drop zone at landing zone STALLION. Attempting to crash land at the landing zone and losing altitude rapidly, the C-130 came under more small arms fire as it tried to turn and crashed and burned at the landing zone.

This tragic crash points out how dedicated our Air Force pilots were during this operation. They were faced with the same miserable weather I have described earlier but, unlike the helicopter, they could not pick holes in the clouds for their descent. They were vectored to the A Shau Valley by the intersection of two radio stations on the east coast. From there, they began an instrument approach into the valley using their own on-board radar to avoid the mountains. No matter how reliable the guages, it takes a lot of guts to poke your airplane nose into clouds that are full of solid rock! Breaking out under a low ceiling, they made their parachute drop and climbed out.

In addition to the superb support of the Hercules pilots, I must mention the 109th Quartermaster Company (Air Drop) at Cam Ranh Bay which rigged 2,212 tons of all classes of supplies for air drop into the A Shau Valley. They and the crews of the C-130's did a tremendous job under extreme pressure.

On the morning of 29 April, elements of the 1st Battalion, 12th Cavalry secured a landing zone in the vicinity of Ta Bat airfield and the 3d Army of the Republic of Vietnam Regiment began its insertion into the valley with the airlift of its 1st Battalion landing at 0830. The 3d Army of the Republic of Vietnam Regiment under the command of Colonel Hoa assumed the security mission for landing zone LUCY. By 1 May the 3d Army of the Republic of Vietnam Regiment had closed on landing zone LUCY and initiated attacks southeast along the Rao Lao River. At 1130 on that date, they discovered a large supply cache of ammunition, vehicle spare parts, and communications equipment. They continued their attack toward the old Special Forces camp at the south end of the valley.

LANDING ZONE STALLION IN THE A SHAU VALLEY, OCCUPIED BY THE 1ST BRIGADE HEADQUARTERS, 1ST CAVALRY

The 8th Engineer Battalion of the 1st Cavalry had been working on the airstrip at A Luoi since 29 April with the heavy equipment that had been brought in by CH–54's. The field was actually ready to accept Caribou aircraft by noon on 1 May. However, it was not until the following day that the first cargo aircraft, a C–7A Caribou, landed at landing zone STALLION at 1120. The engineers continued their work on this airstrip so that it could handle C–123's and C–130's. The 8th Engineers completed the upgrading of the A Luoi Airfield to a C–130 capacity on 3 May.

While the 1st Cavalry Division was operating in the A Shau Valley, the 101st Airborne Division had been conducting major operations to the east. Together with the 3d Army of the Republic of Vietnam Airborne Task Force, they had made periodic contact with the enemy and had uncovered additional major supply caches. The 101st had built up two major fire support bases, VEGHEL and BASTOGNE, which would be important throughout Operation DELAWARE.

After dark on 3 May the 2d Battalion, 3d Army of the Republic of Vietnam Regiment, observed a ten-vehicle convoy approaching their position 7 kilometers southeast of landing zone LUCY. The Army of the Republic of Vietnam unit engaged the convoy with artillery fire, destroying two of the vehicles and causing one large secondary explosion. The rest of the vehicles escaped into the night.

In the next few days, all units engaged in Operation DELAWARE would continue to uncover major enemy supply depots. At the same time, they were exposed to some of the heaviest ground fire received in Vietnam up to that time. Enemy 122-mm rockets were being employed in ever-increasing numbers while smaller caliber shells came in on a much too regular basis.

During this period either I or General Davis worked out of our Division Forward Command Post adjacent to A Luoi Airstrip. This expedited our coordination with non-divisional units such as the 3d Army of the Republic of Vietnam Regiment and enabled us to continue our reconnaissance in force with the greatest flexibility and response. There had been no great changes in our plan after the first assault. Now, planning for a difficult extraction had to be initiated before the monsoon rains became too intense. Our problems now would be to backhaul any excess ammunition and supplies, destroy the maximum amount of enemy supplies, and leave behind thousands of mines and booby traps to make the enemy's future work more difficult. Extraction in many ways proved more difficult than the assault. Rain had already washed out a major portion of the A Luoi dirt airstrip. Consequently, all men and supplies would have to be lifted out by helicopter.

During Operation DELAWARE, our intelligence effort was supplemented by a small Department of Defense team of electronic experts. They came all the way from Washington to emplace the then-new acoustic sensors in the A Shau. The team supplemented our other intelligence capabilities during the operation and placed some sensors that would remain to monitor enemy activity after our forces left the area.

Extraction from the A Shau Valley began on 10 May with elements of the 3d Brigade, 1st Cavalry Division, and the 3d Army of the Republic of Vietnam Regiment being air-lifted out of the valley. The remaining units continued offensive operations in their respective areas, meeting light resistance. The list of captured enemy supplies and equipment continued to grow. A partial list included:

4 track vehicles (including one PT-76 tank)
2 bulldozers
134,757 rounds of small arms ammunition
34,140 rounds of 12.7-mm ammunition
34,332 rounds of 23-mm ammunition
5,850 rounds of 37-mm ammunition
975 rounds of 57-mm ammunition
121 rounds of 75-mm ammunition
229 rounds of 76-mm ammunition
698 rounds of 122-mm gun ammunition
1,680 hand grenades
806 rifle grenades
35 mines
36 mine detectors
2,500 individual weapons
93 crew-served weapons
67 wheeled vehicles (from jeep to 2½-ton truck)
2,182 pounds of explosive
5,994 blasting caps
31 flame-throwers
90,000 pages of documents
71,805 pounds of food stores.

Operation DELAWARE-LAM SON 216 was officially terminated on 17 May. General Rosson labeled Operation DELAWARE as

> one of the most audacious, skillfully executed and successful combat undertakings of the Vietnam War . . . It is significant that from its inception DELAWARE was a combined effort entailing association of the 1st Cavalry Division and the 3d ARVN Regiment, 1st ARVN Division, on the one hand, and the 101st Airborne Division and the 3d ARVN Airborne Task Force on the other. The outstanding results achieved through teamwork on the part of these combined forces reflect great credit on their leadership, professionalism, and unsurpassed fighting zeal.

The 1st Cavalry Division had gone into the A Shau Valley in the face of the heaviest enemy air defense ever encountered in airmobile operations up to that time. According to the military textbooks, this would have been called "medium intensity" antiaircraft fire. To the pilots who had to fly into this area day in and day out, "medium" hardly seemed the appropriate descriptive word. There is no doubt that the enemy had ringed this important base with sophisticated heavy antiaircraft defenses including 37-mm weapons. They had planned and interlocked their fire zones over a period of some years. While the 1st Cavalry Division lost twenty-one helicopters in this operation, the fact that they were able to make a major move into such an area in the face of this threat and under the worst possible weather conditions is a tribute of the soundness of the airmobile concept. Some of the helicopters that were lost ignored clear warnings of intense enemy concentrations that had been uncovered by prior reconnaissance. At times the weather gave an additional aid to the enemy by channelling helicopters into certain flight paths to go underneath the clouds. The enemy, of course, adjusted his fire to the obvious approaches.

The list of ammunition, weapons, and equipment that was destroyed in Operation DELAWARE was most impressive in itself. This was, in a manner of speaking, the enemy's Cam Ranh Bay. However, I believe one of the greatest intangible results of this operation was the psychological blow to the enemy in discovering that there was no place in Vietnam where he could really establish a secure sanctuary. The enemy had always considered A Shau Valley to be his personal real estate and it was a symbol of his relative invulnerability. Operation DELAWARE destroyed that symbol.

From the allied point of view, Operation DELAWARE brought out one important consideration. Weather had been the key planning factor on the timing of this operation from the beginning. The urgency to terminate Operation PEGASUS in order to go into the A Shau Valley was based on inches of rain to be expected after the month of April, not *ceilings* and *visibilities* which would prove so critical. In other words, the forecast monsoon rains (which did occur) never produced the terrible flying conditions of low ceilings and scud which preceded them in April. An air cavalry division can operate in and around the scattered monsoon storms and cope with the occasional heavy cloudbursts far better than it can operate in extremely low ceilings and fog. The monsoon rains did, in fact, wash out the hastily constructed airfield but our capability for airmobile operations improved during the period. The lesson learned, then, was that one must be very careful to pick the proper weather indices in selecting an appropriate time for an airmobile operation. An inch of rain that falls in thirty minutes is not nearly as important as a tenth of an inch which falls as a light mist over 24 hours. According to the long range forecast based on old French records, April was supposed to have been the best month for weather in the A Shau Valley. As it turned out, May would have been a far better month—but you don't win them all.

CHAPTER X

Airmobile Developments, 1968

Change of Command at Military Assistance Command, Vietnam

On 11 June 1968, General William C. Westmoreland passed command of the U. S. forces in Vietnam to General Abrams, who would serve as Acting Commander until General Westmoreland was sworn in as Chief of Staff of the Army on 3 July.

It would be presumptuous on my part to try to give an overall assessment of General Westmoreland's influence on the airmobile concept during his four and one-half years as Commander, U. S. Military Assistance Command, Vietnam; however, there was no doubt that General Westmoreland had set the airmobile stage for his subordinate commanders through his own strategy and example. Most importantly, after giving them his broad concept of an operation, he allowed them that precious freedom which is so necessary for subordinates if they are expected to exercise their own imagination and seize fleeting opportunity.

For this study, it is important to again note that General Westmoreland believed so much in the potential of airmobility that he was willing to commit the 1st Cavalry to a major operation only days after its arrival—recognizing full well that a failure would have been not only a setback for the war but a disaster for the future of airmobility. Again, his decision to relieve Khe Sanh by a series of air assaults under the scrutiny of the entire world press was solid affirmation of his confidence in the airmobile concept. Appropriately, he would become the first Army Chief of Staff to wear the Army Aviator Badge.

Enemy Helicopters?

On the night of 16 June 1968, an incident occurred that graphically demonstrated the problems of command and control of the air space in the face of an enemy air threat. It also surfaced fundamental differences in the Air Force and Army concepts which, if not solved, could have a major impact on future airmobile tactics.

At 2055 hours on 16 June a U. S. radar station reported that ten unidentified helicopters had been located six kilometers north of the Ben Hai River. During the remainder of the night of 16–17 June 1968 numerous reports were received of enemy helicopters operating in the vicinity of the Demilitarized Zone. Another report cited an attack on a U. S. Navy boat by an unidentified aircraft. Finally, it was reported that many of the enemy helicopters were destroyed by U. S. Air Force aircraft and by artillery. Two hours after the first report had been received, 7th Air Force dispatched a message stating that all aircraft, both helicopter and fixed-wing, operating in the I Corps Tactical Zone would be under the positive control of Air Force ground radar stations. In effect, this Air Force message would have paralyzed all ground operations in this area.

General Williams, Commanding General of the 1st Aviation Brigade, visited 7th Air Force Headquarters early on the morning of 17 June to inform the Chief of Staff of the 7th Air Force that compliance with the Air Force teletypewriter exchange was impractical since it would cut off not only tactical operations but also emergency resupply and medical evacuation flights in I Corps Tactical Zone. He pointed out that the 7th Air Force did not have a control means to handle even a fraction of the 1,000 Army aircraft that were in I Corps and that the 7th Air Force message itself was not valid since that headquarters was not in the U. S. Army, Vietnam, command channel. There is no record of III Marine Amphibious Force receiving the message or, if received, any objection to the message. At 1100 on 17 June the Chief of Staff of Military Assistance Command, Vietnam, informed the Chief of Staff of the 7th Air Force that Military Assistance Command, Vietnam, had not even been an information addressee on the message; that the message was impractical since it would stop combat operations; that it was not within the prerogative of the 7th Air Force to send; and that it would be rescinded.

Colonel Eugene M. Lynch, Deputy Aviation Officer, U. S. Army, Vietnam, attended a meeting at 7th Air Force Headquarters that same day to discuss this problem. The Air Force considered the helicopter sightings a serious threat to Da Nang Air Base and all other major installations in the I Corps Tactical Zone. Colonel Lynch emphasized the points that had been made previously . . . that combat operations could not be stopped entirely and that the Army installations were quite capable of defending themselves against enemy helicopters. He suggested that the best anti-enemy helicopter weapon might be the armed Cobra. The Air Force continued to press for drastic measures to cope with the threat.

A meeting was held at III Marine Amphibious Force on 18 June to discuss the Services positions and problems in this matter. The Marines and the Army took similar positions, basically that stated previously by General Williams. The Air Force cited one of their problems was that of identifying the low slow-flying aircraft.

As time passed and further investigations were conducted, it developed that no finite evidence was available to confirm that enemy helicopters had been observed in flight, on the ground, or in a damaged state following claims of destruction by friendly air. The previously accepted positive reports were discredited.

My purpose in relating this incident is not to criticize any Service or persons involved, but to highlight the lack of prior planning in this most important area. The U. S. forces had enjoyed complete air superiority for so many years in Vietnam that the mere possibility of the presence of even an enemy helicopter threw all of their operational procedures into chaos. The Air Force reaction to a possible enemy air threat was predictable and, in their estimation, the only realistic option at that time. The Army and the Marine ground elements, which depended on the unlimited use of the helicopters, naturally could not accept the rigid restrictions that the Air Force insisted was necessary. The incident did not have a positive effect of clarifying the command channels and some sober planning on means and methods to better react to future enemy air contingencies. It is an incident that should not be forgotten by the future planners of airmobility in any conflict where absolute air superiority is not a basic assumption.

The Second Airmobile Division

On 28 June 1968 U. S. Army, Pacific published General Order 325 which initiated reorganization of the 101st Airborne Division into the Army's second airmobile division. This same order called for the Division to be redesignated the 101st Air Cavalry Division effective 1 July 1968.[1]

[1] At this same time, the 1st Cavalry Division was redesignated the 1st Air Cavalry Division. Indeed, the abbreviation "1 ACD" had been in common use long before this official change. However, the terminology "air cavalry division" was revoked by Department of the Army on 26 August 1968, and the designations were reestablished as the 101st Airborne Division (Airmobile) and the 1st Cavalry Division (Airmobile). To avoid confusion, this monograph only refers to the two divisions' latter designations.

Conversion of the 101st to an airmobile configuration had been considered by Department of the Army prior to the deployment of the division (–) in December 1967. However, the continued deficit in aviation assets during the buildup of forces in Vietnam had made such conversion impractical. The deployment of the 1st Brigade of the 101st in 1965 has been covered in an earlier portion of this study. In August 1967, two years and twenty-seven days after the departure of the 1st Brigade, the remainder of the 101st Airborne Division was alerted for deployment to Vietnam from Fort Campbell, Kentucky. Originally, the Division was scheduled to be in country on 10 February 1968. However, this movement was subsequently rescheduled for early December 1967.

The culmination of the Division's preparation for deployment was Operation EAGLE THRUST. This lift was keynoted by the departure on 8 December of the Commanding General, Major General Olinto M. Barsanti, in an aircraft piloted by General Howell M. Estes, Jr., Commander of the Military Airlift Command. On 18 December 1967 the last airplane touched down in Vietnam ending the largest and longest military airlift ever attempted into a combat zone. The move had required 369 C–141 Starlifter aircraft missions and 22 C–133 Cargomaster aircraft missions, ultimately airlifting 10,024 troops and over 5,300 tons of the Division's essential equipment.

During the 1968 Tet Offensive, elements of the Division moved to protect the cities of Saigon, Bien Hoa, Song Be, Phan Thiet, and Hue. As mentioned earlier, elements in Saigon had helped secure the American Embassy during the first few hours of that abortive enemy attack. The Division Headquarters moved to the Hue area on 8 March 1968. The 101st would remain in this area for the next few years.

The Screaming Eagles participated in a series of combat operations in the I Corps Tactical Zone to include Operation DELAWARE near the A Shau Valley. Though the 101st was not programmed into the A Shau Valley proper during this operation, it would be back in force several times in the next few months. When the order came to reorganize as an airmobile division, the 101st was involved in Operation NEVADA EAGLE, a large rice-denial effort in the plains south of Hue.

The Division developed a three-phase plan to accomplish the conversion from the airborne to the airmobile configuration. The first phase (1 July to 1 December 1968) would involve the activation and organization of the 160th Aviation Group and a reorganization of the division base. The second phase, which would not be completed until June 1969, involved the conversion of the armored cavalry squadron to an air cavalry squadron. The last phase involved the activation of an aerial rocket artillery battalion. It was determined that a full year would be necessary to convert the 101st to an airmobile configuration. This time lag took into consideration the long range procurement of aviation assets and the fact that the Division would continue to conduct combat operations throughout the conversion period without degradation to its combat posture.

The reorganization progressed in a smooth manner with a few exceptions, one of these being in the aircraft maintenance area. U. S. Army, Vietnam, and the 101st planners envisioned the new airmobile division's aircraft maintenance as decentralized rather than centralized. In this case, they hoped to profit from the experience of the 1st Cavalry Division and the 1st Aviation Brigade. Higher headquarters had failed to establish this cellular concept in its order for the conversion. Instead, it called for a large centralized maintenance and supply battalion. This problem was resolved by U. S. Army Pacific General Order 607, dated 11 October 1968, which established the desired cellular maintenance organization within the division. In effect, this married a maintenance cell with each company-sized aviation unit. This cell would have the capability of providing both direct support and avionics maintenance.

Many "chicken and the egg" arguments resulted from the final maintenance structure of the 101st versus the then current maintenance structure of the 1st Cavalry Division. The 1st Cavalry had espoused decentralized maintenance ever since its deployment to Vietnam, but there was finite limitations on the number of skilled mechanics and maintenance tool sets during the early build-up phase. The arguments of the logisticians for centralized maintenance prevailed.

When I took over the division in 1967, the division was still tied to the maintenance "hub" at An Khe. Only the detached battalion combat team at Phan Thiet had its own maintenance capability. Operation LEJEUNE, a week after I joined the division, convinced me that drastic action would have to be taken, and I ordered our logistics planners to develop maintenance kits that could be easily moved with displaced units. In October 1967, when I had to send a brigade to Chu Lai, the problem became critical. Part of a maintenance battalion had gone with the brigade, but there was still insufficient capability.

U. S. Army, Vietnam, in the early planning for the move to I Corps Tactical Zone, finally recognized our obvious requirement, and included a more decentralized maintenance concept as part of their logistical plan. I've already mentioned how fortunate we were to have started our logistical bases at Red Beach and Hue-Phu Bai prior to the *Tet* Offensive. When the 1st Brigade went to Quang Tri, they *had* their maintenance with them. We had finally broken down the "golf course" syndrome which was associated with An Khe.

When the 101st was ordered to an airmobile configuration, my staff and I completely supported its concept of decentralized maintenance. The major portion of the tremendous responsibility for converting the 101st to its new configuration would soon be in the very capable hands of Major General Melvin Zais. He would receive the unqualified cooperation of every aviation commander in this effort.

Thoughts on Leaving the Cavalry

On 15 July 1968 I turned over temporary command of the 1st Cavalry Division to Brigadier General Richard L. Irby. Major General George I. Forsythe had already been announced as the new Division Commanding General and would formally take command on 19 August, after completing flight training. After fifteen months of combat, several points stood out in my mind as particularly significant to the airmobility concept.

First, the vulnerability of the helicopter was still the most debatable issue in the entire military establishment. Depending on one's preconceived notions, one could use the same statistics to prove opposite viewpoints. For example, during calendar year 1967 the 1st Cavalry had flown 977,983 sorties and had 688 aircraft hit. Of these, 36 were shot down and lost. In the first six months of 1968, the Division had flown 407,806 sorties and had had 271 aircraft hit. Of these, 66 were shot down and lost. Part of this difference could be attributed to the increased intensity of combat in the northern I Corps Tactical Zone where the Division had been fighting since January; part could be attributed to the improvement in the quantity and quality of the North Vietnamese Army antiaircraft weapons systems; but, much of this so-called "increased vulnerability" was due to the thousands of random and sometimes uncontrolled flights by one or two aircraft performing separate observation, administrative, and control missions. In other words, I do not believe that we had a major increase in our losses in our organized air assault formations, even during the high intensity antiaircraft fire of the A Shau Valley. However, more and more helicopters were being flown singly and in pairs without the benefit of detailed pre-flight intelligence briefings on especially "hot" zones and known enemy concentrations. It was particularly hazardous for the many transient aircraft that assumed they were in "safe air" just because they were a few kilometers from a major U. S. installation. Before I left the Division, I initiated a formal study effort to quantify this observation. Unfortunately, this study was never completed.

The second major observation I made in my after action report of 15 July 1968 was the critical problem of protecting the airmobile division's helicopters on the ground. Until one has viewed at first hand the real estate necessary to park approximately 450 dispersed helicopters near their related aircraft maintenance facilities, it is difficult to appreciate the magnitude of this problem. I know that the Marines grossly underestimated our base camp requirements when we moved north into the I Corps Tactical Zone. They just could not believe that we really required the area we had asked for—the equivalent of several major airfield complexes.

Once established, these areas had to be secured. This security was partly obtained by dispersion, construction of revetments, and organization within the landing zone to separate the helicopters from ammunition and fuel storage areas. Security was further enhanced by armed helicopter patrols throughout the night and roving patrols on the ground. The ground patrols had to operate at least to the range of enemy mortar. The security requirements sometimes caused a tremendous drain on combat resources if they were not monitored closely by each tactical commander. However, it was my experience that an airmobile division cannot overlook any detail in the security of its most valuable asset—its helicopters.

My third comment concerned unit integrity. I've described earlier in this study how the 1st Cavalry Division was often "fragmented" to send forces of battalion size and brigade size to other areas of operation. On the surface, it might appear likely that the division could have easily sent a proportionate share of our aviation assets and maintenance to the same operation. As a matter of fact, there was no proportionate share of airlift that automatically went with any brigade or battalion in the division. The division was conceived so that the division commander could vary the amounts of *his* assets that would be furnished to any portion of *his* force during an operation. There was no set quota of airlift that was allocated on a daily basis to any unit in the division. When the division was fragmented it reduced the flexibility of the division as a whole and proportionately reduced the true capabilities of the smaller forces. This is not to say that the airmobile division cannot be used as the "fire brigade" for a theater commander; however, like the air cavalry squadron, its total effectiveness can only be realized as an entity.

My fourth observation was the constant need to be alert to the over-utilization of rotary wing aircraft in the supply mission. While I was the commander, I emphasized that our planning must always include provisions for utilizing the Caribou, C–123, and C–130 aircraft wherever they could enhance our capabilities. I knew of no instance during my tour in Vietnam when the 1st Cavalry Division was denied this fixed-wing support when it was requested. There were occasions when the division staff failed to plan for these aircraft and took the easy way out by going ahead and using our organic Chinooks for long logistics hauls. Operation PEGASUS was a good example of careful planning for the maximum use of the fixed-wing capabilities in the operational area. It would have been impossible to move the tonnages required in the time necessary without the careful preparation of landing zone STUD as a forward fixed-wing base.

As a fifth observation, I expressed my disappointment in one item of equipment, the Mohawk. From the Division's point of view, and I emphasize *division*, the six organic Mohawks did not provide the expected intelligence. We made a sincere and continuous effort to use the Mohawk's side-looking airborne radar capability and the infra-red capability, but the end product never justified the man-hours and assets required. I was particularly disappointed in these results, because I had spent so much effort as Deputy Director and later as Director of Army Aviation in justifying this system. In the early years we had made a big point of the necessity of having this capability at the division level. We had hoped that the division commander would have a "real-time" read out of valuable information through a data link. After fifteen months, I had to conclude that the Mohawk should be employed at corps and theater level.

Finally, no summary of my experience with the 1st Cavalry Division would be complete without mentioning once again the "man" part of the "man-machine" equation of airmobility. During this period, I was particularly fortunate to have a series of outstanding officers and warrant officers. Space has not permitted mentioning even a small percentage of their names. But, equally important, the 1st Cavalry Division was blessed with a continuous flow of courageous young men in our noncommissioned officer and enlisted ranks. My notes of 15 July 1968 read like this:

I have complete confidence in our rapid production of NCOs which we must continuously produce here within the Division. These men are smarter than the ones we had in World War II. They are just as gallant and courageous also. The ones that have the talent must be spotted early and must be promoted just as fast as it is possible to do so. They do a tremendous job and respond to the challenge immediately. Continuous checks must be made to see that units are promoting men who are doing the job just as fast as possible. I do not think there should be any great concern of the caliber of NCOs that we have today in Vietnam because of their youth and brief experience. If they are carefully selected and assisted as much as possible, they will carry the ball. In fact the young soldier that we have in this Division today is the greatest our Army has ever had during my service. There is a wealth of material ready to become competent combat leaders.

Status of the 1st Aviation Brigade

In July of 1968 the 1st Aviation Brigade reassigned the 308th Aviation Battalion together with its support detachments to the 101st Airborne Division. This was part of the in-theater readjustment necessary for the airmobile reconfiguration of this division. This reassignment still left the 1st Aviation Brigade as the largest single Army aviation command in the world; indeed, larger than the air force of most countries. As of 31 July 1968, its strength was 25,181 men distributed as follows:

	Officers and Warrant Officers	*Enlisted Men*
Headquarters and Headquarters Company, 1st Aviation Brigade	61	204
12th Aviation Group	1644	6828
16th Aviation Group	456	2883
17th Aviation Group	1468	6577
164th Aviation Group	593	3065
58th Aviation Battalion	84	1228
TOTAL	4306	20875

The chronic shortage of aviation personnel described in earlier chapters had finally been overcome by the expanded output of the Army Aviation School and the leveling off of requirements in Vietnam. Aviators, once frozen to their cockpit positions, were again receiving normal ground assignments and schooling. Though many aviators were still forced to anticipate recurring tours in

Vietnam, they could count on more time between such tours and more diversified assignments to fill out their career pattern.

In Southeast Asia, the Army aviator had become the *sine qua non* of combat operations. No major plan was ever considered without first determining the aviation assets available to support it. Nowhere was this better exemplified than in the 1st Aviation Brigade.

On the second anniversary of this unit back on May 25th, General Abrams, Deputy Commander, U. S. Military Assistance Command, Vietnam, summed up the feeling of the non-rated officers this way: "It has always been interesting for me to note that the aviators and men of this Brigade have been taken into the brotherhood of the combat arms. Not by regulation, not by politics, but they have been voted in by the infantry, who are the chartered members of that secluded club, the combat arms." During this same organizational day ceremony, noting the presence of General Cao Van Vien, Chairman of the Joint General Staff, General Abrams added, "They are heroes to the district chiefs; they are heroes to the province chiefs; and they are heroes to soldiers of every nation that fights here."

During the same ceremony, General Vien presented the Brigade its second Vietnamese Cross of Gallantry with Palm, an award earned by the men of the Brigade for their outstanding aerial support of such operations as JUNCTION CITY and JEB STUART, and their opposition to the *Tet* offensive. General Williams, the Commanding General of the 1st Aviation Brigade, listed a few of the Brigade's accomplishments during 1967 when they airlifted more than five million troops—the equivalent of 313 infantry divisions—in more than 2.9 million sorties. In that year Brigade aircraft flew more than 1.2 million hours—the equivalent of 137 years. The Brigade was credited for killing 10,556 Viet Cong, sinking nearly 10,000 supply sampans, and destroying more than 10,400 enemy structures and fortifications.

It is very difficult to properly document the accomplishments of the pilots and crews of the 1st Aviation Brigade since their deeds have been interwoven in the combat operational reports of the units which they supported. This support almost became accepted as routine.

I have tried to spare the reader the series of inevitable "wiring diagrams" so beloved by many students of military organizations. However, the organizational structure of the Brigade as of 31 July has a special impact in the sheer number of separate aviation units that were supporting the Free World Forces at this time. (*Chart 2*)

Chart 2—1st Aviation Brigade Organization, 1 August 1968

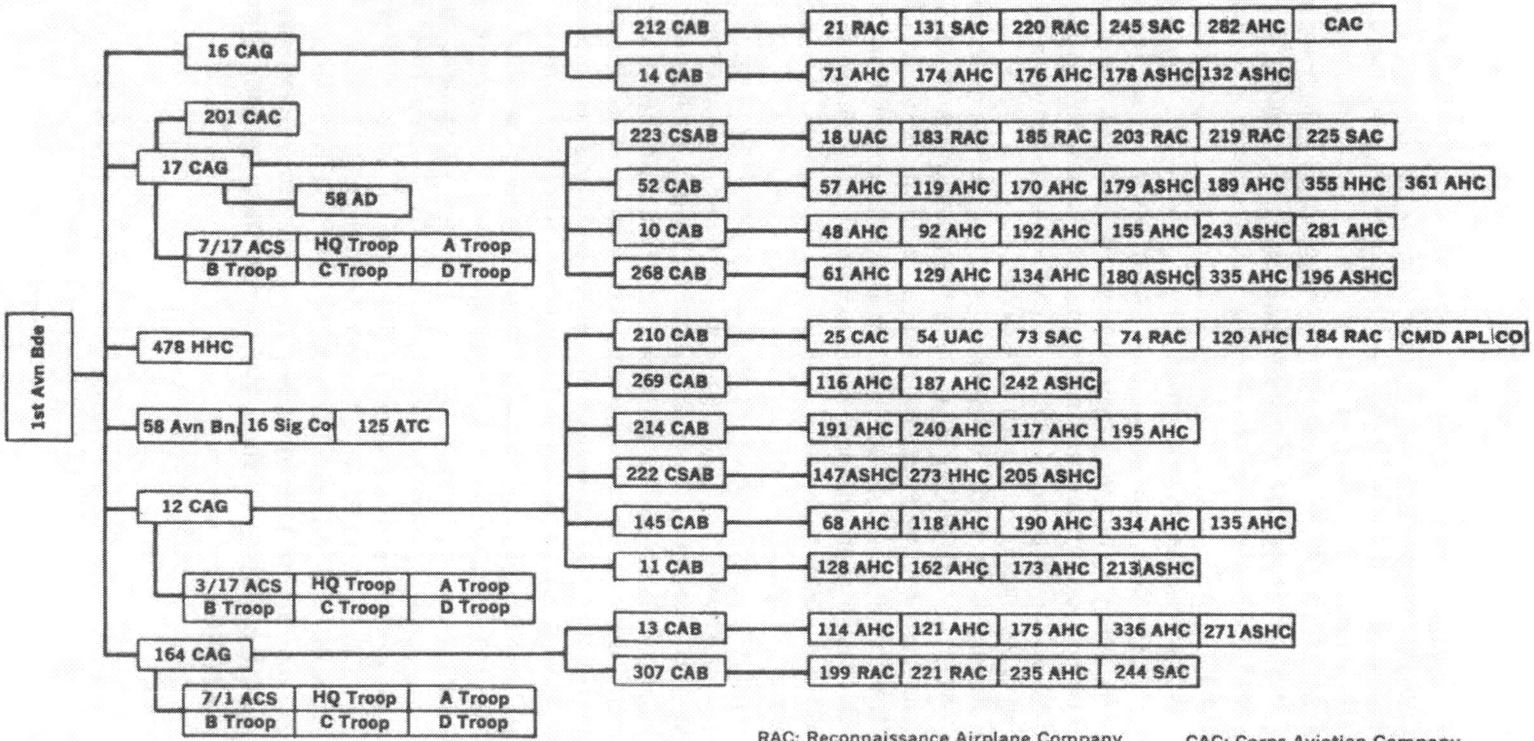

Key

CAG: Combat Aviation Group
CAB: Combat Aviation Battalion
AHC: Assault Helicopter Company
UAC: Utility Airplane Company
HHC: Heavy Helicopter Company
RAC: Reconnaissance Airplane Company
ASHC: Assault Support Helicopter Company
AD: Aviation Detachment
CSAB: Combat Support Aviation Battalion
CAC: Corps Aviation Company
SAC: Surveillance Airplane Company
ACS: Air Cavalry Squadron
ATC: Air Traffic Control Detachments

The Brigade was not only continuously supporting combat operations throughout the length of Vietnam, but also expanded its operation of in-country schools for the training of Vietnam Air Force Huey pilots, Republic of Korea and Thai aviators, as well as U. S. and Australian personnel.[2] The Brigade was also responsible for the supervision of the New Equipment Training Teams which eased the introduction of complex equipment into the operating units.

These training teams performed a vital function in smoothing the way for complex equipment that was introduced rapidly into combat. A small group of experts would arrive with every new aircraft system and armament sub-system and work with in-country personnel to build up experience and iron out maintenance problems that are inherent in any new system.

For example, back in September 1967 when the first Huey Cobra (AH-1G) arrived in Vietnam, a New Equipment Training Team was ready for the aircraft. This team used the first six Cobras to check out the pilots of the 334th Assault Helicopter Company. When the design problems and maintenance problems were solved, these Cobras were then assigned to that company. The New Equipment Training Team pilots traveled from unit to unit to standardize procedures and to record early problems which were immediately relayed to Continental U. S. Some of the problems were not technical but more operational in nature, such as experienced in the Cobra where pilots missed the telltale noises of enemy fire, which they could hear through the open door of the older armed Huey, and the additional eyes and ears of the door gunners. Some modifications of tactics and techniques were necessary.

The New Equipment Training Teams highlight another facet of the growth of airmobility in Vietnam. Never in any previous war had there been so much new and complex equipment introduced into actual combat in such a short period with a minimum amount of problems. The transition from the old fleet of worn out H-21's to a modern turbinized sophisticated fleet of thousands had been accomplished on the battlefield. The combination of the training base in the United States, an aerial line of communications, superb logistics, and the know-how of American industry had made this possible.

---

[2] One unique company of the Brigade, which supported the Australian Task Force, was composed of 50 percent Australian Navy personnel with the other half U. S. Army.

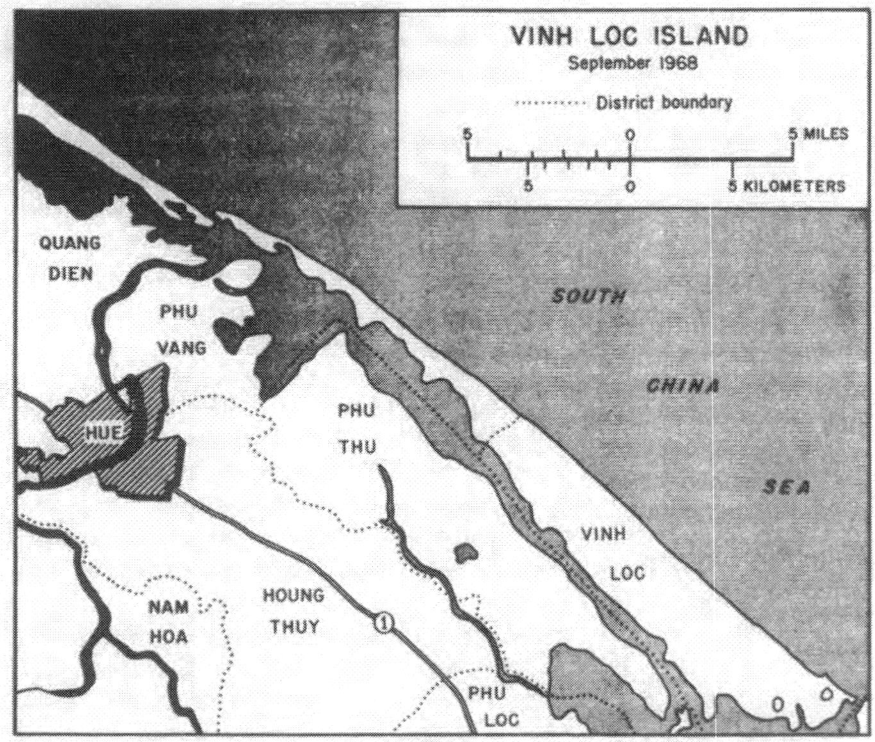

MAP 8

An Example of Cordon Operations

Airmobile forces often can be best employed with other forces to achieve unique capabilities when a special opportunity presents itself. I have previously described airmobile operations with armor in Binh Dinh Province. The following operation, in which the 101st Airborne Division participated in a combined air-sea-land maneuver, demonstrates again the imaginative use of all Free World capabilities. Helicopters, air-cushion vehicles, swift boats, and armored personnel vehicles were effectively employed in an integrated force.[3] But Operation VINH LOC was not a mere demon-

[3] The air-cushion vehicles used their unique capability of operating in swampy or very shallow water to patrol otherwise inaccessible areas. Most of these machines saw duty in the Mekong Delta region where they had done a magnificent job. I have not treated them extensively in this study, since they are not really an extension of airmobile operations so much as riverine operations. During the 1950's, the Army experimented with a variety of so-called "ground effects machines." They were more noted for their instability and huge "signature," than their practicality.

stration of modern technology and sheer power; it was a microcosm of the problem facing all our forces in Vietnam—the shadow Viet Cong "government" and how to deal with it.

Operation VINH LOC, a combined operation of elements of the 101st Airborne Division, the U. S. Navy, and Republic of Vietnam forces was conducted in Thua Thien Province during the period 10–20 September 1968. Vinh Loc Island, fifteen miles east of Hue, is a sub-tropical island with sandy beaches, palm trees, and rice paddies. (*Map 8*) It is almost completely flat with the exception of two small hills on the southeastern end.

The *Tet* offensive had left a power vacuum on the island since most of the Regional and Popular Force units had been pressed into service at Hue. Viet Cong forces moved into this vacuum and established a sanctuary, prepared defensive positions, and established caches. As a symbol of their power, they had staged a public execution of a village elder and a thirteen-year-old girl to demonstrate the consequences which local residents could expect if they failed to support the Viet Cong.

One allied operation had been conducted on the island prior to Operation VINH LOC. On 9 July 1968 the 2d Battalion, 7th Marines, 1st U. S. Marine Division, had begun a one-week operation. The Marines did not encounter organized resistance, for most of the enemy preferred to hide or mingle with the civilians rather than fight. The civilians were reluctant to provide any information about the enemy because they felt the Viet Cong would reassert their dominance as soon as the Marines withdrew.

When three Regional Force companies tried a sweep operation of the area on 4 September, they were pinned down by intense mortar, rocket, and small arms fire. Only the suppressive fire of the armed helicopters of the 101st Airborne Division allowed these units to break contact. American advisors estimated that there were at least two Viet Cong reinforced companies and possibly a battalion in the area.

While these activities were occurring, plans were being made for a combined operation to end enemy influence on the island. The concept of operations envisaged cordoning the island with naval and ground forces while air assaults were planned to overcome organized resistance and to fragment enemy units. All available district resources would be used in this search.

General Zais, Commanding General, 101st Airborne Division (Airmobile), Major General Truong, Commanding General, 1st Army of the Republic of Vietnam Division, and Colonel Le Van Than, Province Chief of Thua Thien, approved of the concept for a cordon and search of the portion of Vinh Loc Island which comprised Vinh Loc District. Final coordination was completed on the 9th and 10th of September.

An Area Coordination Center was established under the overall direction of the Commanding Officer, 2d Brigade, 101st Airborne Division. This center was staffed by the Deputy Province Chief, the U. S. Deputy Sector Advisor, the District Chief, and the Commanding Officer, 1st Battalion (Airborne), 501st Infantry. The center also included an intelligence section comprised of representatives from the province and district intelligence staffs, the U. S. battalion S-2 section, Special Branch Police, National Police, and the Census Grievance Committee. The entire group would be moved to Vinh Loc District Headquarters after the combat assault.

On 9 September one Regional Force company withdrew from the island as planned. The next day Company D, 1st Battalion (Airborne), 501st Infantry moved from the northeastern end of the island to an area just north of the Vinh Loc-Phu Vang District boundary. The paratroopers had been operating in this area to provide security for the Col Co beach naval supply facility, so this maneuver did not arouse enemy suspicion. It was also designed to drive enemy forces southeast along the island into Vinh Loc District. The 54th Army of the Republic of Vietnam Regiment had been conducting operations in Phu Thu District. The 2d and 3d battalions of the regiment moved into blocking positions along the northeast coast of Phu Thu District to drive the enemy across the water into Vinh Loc. This maneuver was also an extension of normal operations and did not reveal to the enemy the impending move into Vinh Loc. The 3d Troop, 7th Army of the Republic of Vietnam Cavalry moved with its armored personnel carriers to the Hue Landing Craft, Utility, ramp, then down the Perfume River on medium landing craft to the Tan My LST ramp opposite Col Co beach. They remained in the Tan My area until the night of 10 September and, under the cover of darkness, crossed over to Vinh Loc to a position south of the Col Co beach facility.

During the night of 10–11 September, the U. S. and Vietnamese Navy converged around the island and at first light on 11 September the encirclement of the island was complete. At 0732 three companies of the 1st Battalion, 501st Infantry conducted air assaults into three landing zones on the seaward side of the island and started reconnaissance in force to the south. Forty-five members of the National Police Field Forces had been evenly distributed between the three companies. The 1st Battalion of the 54th Army of the Republic of Vietnam Regiment conducted an air assault into three landing zones at the same time to the northwest of the 1st Battalion, 501st Infantry. Popular Force platoons joined each of the rifle companies after the combat assaults.

All personnel encountered were detained for questioning by the Special Branch Police. These police interrogated detainees immediately and were able to classify innocent civilians on the spot. The latter were then escorted to safe areas. All other detainees were evacuated to the combined intelligence center. Here those classified as Viet Cong or North Vietnamese soldiers were moved to a Prisoner of War compound and evacuated through U. S. channels after interrogation. Members of the Viet Cong infrastructure and civilian dependents were evacuated through province channels. Hamlet and village chiefs were used to assist in the identification.

An interesting ruse was perpetrated on the enemy by the district S–2. As darkness approached on 11 September, the Army of the Republic of Vietnam cavalry troop and infantry battalion had 231 detainees to be evacuated to the combined intelligence center. Beginning at 2220 hours, CH–47 helicopters arrived to transport these personnel to the center. As the first lift landed and the detainees began disembarking, the district S–2 was waiting for them. Amid the dust and confusion, he commanded in a loud voice, "K4B (battalion) over here, C117 (company) over there, C118 (company) over there." To the surprise of everyone, except perhaps the S–2, sixty-three of the detainees segregated themselves as he had directed. The S–2 then took personnel from those who had obeyed his command to those who had not responded and asked, "Is there anyone from your unit who did not fall out?" Additional prisoners were thus identified.

Numerous Eagle Flights were used to conduct ambushes and to rapidly exploit intelligence as it developed. Each company conducted detailed searches in its zone. While ground forces were ferreting out the enemy, the Navy and the 2d and 3d Battalions, 54th Army of the Republic of Vietnam Regiment, maintained an effective cordon of the island. The results of this relatively small operation, which ended on the 20th of September, were very impressive. One hundred and fifty-four enemy were killed, 370 were captured (including 116 members of the infrastructure), and 56 Viet Cong rallied to the government cause. Only two friendly troops were killed and nine wounded.

Equally important, it should be noted that the destruction of civilian property and the danger to the civilian populace were held to an absolute minimum. Preparation fires for the combat assaults were limited to barren areas near the landing zones. Though supporting fires were available throughout the operation, only two fire support missions were necessary. As an indication of the restraint placed on the destructive power available, only two civilians were wounded and three huts destroyed during the entire operation.

The success of Operation VINH LOC was due to its detailed planning and coordination and the full utilization of all resources available—airmobile, naval, and ground. The establishment of the combined command post made possible the immediate flow of information between intelligence, operations, and fire support elements. The integration of the Popular Force platoons and the police forces into the U. S. units overcame the language barrier and permitted rapid distinction between civilians and enemy forces. Most importantly, the island was not left in a vacuum when the operation was over. One U. S. rifle company and an Army of the Republic of Vietnam battalion remained on the island to assist in the resettlement of the population, and to aid district forces in maintaining a secure atmosphere. Self-defense cadre were trained at district level, and rural development teams were inserted to assist in the re-establishment of local government, and to help the people improve health and sanitation measures, as well as agricultural methods. Operation VINH LOC, although modest in size, had shown how the United States and Republic of Vietnam forces could enhance their individual capabilities through mutual planning, coordination, and cooperation.

The Cavalry Moves South

One of the more difficult military tasks is to move a division in contact with the enemy to another area of a combat theater, reposition it quickly, and have it ready to fight. In Vietnam where there were never any front lines nor truly secure rear areas, such a move had special problems. Operation LIBERTY CANYON, which was the code name given to the move of the 1st Cavalry Division from I Corps Tactical Zone to III Corps Tactical Zone in the fall of 1968, is one of the best examples of rapidly moving an entire division—a move that was made so professionally and smoothly that it even achieved strategic and tactical surprise to a knowledgeable enemy. The 1st Cavalry Division had become the world's acknowledged expert at pulling up stakes.

General Forsythe received a phone call on the night of 26 October 1968 alerting him that his division would move as quickly as possible by air and sea to the III Corps Tactical Zone and take over a new area of operation. This proposed move, which was in response to a growing threat in the south, meant that the remaining forces in I Corps Tactical Zone would have to rapidly readjust their positions to assume the large area of responsibility of the 1st Cavalry Division so that no vacuum would be left into which the enemy could infiltrate. Speed was of the essence and planning time would be at a minimum.

On 27 October General Forsythe, accompanied by four key staff members, departed Camp Evans and arrived at Headquarters II Field Force Vietnam at approximately 2230. The group was briefed by their equivalent members of the staff of Lieutenant General Walter T. Kerwin, Jr., Commanding General, II Field Force, Vietnam. The initial meeting determined that the move would begin on the 28th by air and sea with one brigade closing by 2 November. The 1st Cavalry staff members furnished II Field Force Vietnam planners the division's estimated daily consumption of supplies and fuel, and estimated exactly when and where these supplies would be needed during and after the move. Aircraft density by type and ground equipment density by type had to be determined as of the time of move and hasty decisions made as to what would go by air and what would go by sea. A freeze was established on all major personnel changes until further notice.

On the following day the groups met with Lieutenant General Frank T. Mildren, Deputy Commanding General, U. S. Army Vietnam, and members of his staff. During this meeting, important decisions were made concerning the disposition of the aviation units, engineer effort for revetments and storage facilities, and time phasing for the brigades of the Division to begin operations in their new area. Also, it was determined that the Division base would be located at Phuoc Vinh and the Division Support Headquarters would be located at Bien Hoa. A temporary Forward Command Post would be established at Long Binh for a period not to exceed two weeks.

Many different agencies and all Services had to participate in the planning at a very early stage. The Air Force would have to know the number of C–130 sorties required against the maximum sorties per day available. The Navy needed to know the size and quantity of equipment to be moved by sea and where it would embark and debark. They would then determine the shipping that could be made available for the move. The Movement Control Center, Saigon Support Command, the 1st Logistical Command, and the 1st Aviation Brigade were brought into the planning during the next day.

While these meetings were taking place in III Corps Tactical Zone, similar meetings were taking place in I Corps Tactical Zone at Headquarters III Marine Amphibious Force. Here it was determined that three LST's would be available for loading at Tan My on 28 October and two LST's available each day thereafter until completion of the move. The aircraft carrier *Princeton* was made available to move aircraft and equipment from the Da Nang area. Transportation was set up to haul organizational equipment from division unit locations at landing zone BETTY-SHARON, landing zone NANCY, and Camp Evans to loading facilities at Tan My Ramp, Quang Tri Airfield, and Phu Bai Airfield.

All the units of the division had contingency plans for air movement but these had to be modified on the spot to separate what was to be moved by air and what would be moved by sea. At Camp Evans, General Irby, Assistant Division Commander Alpha, directed that 1/4-ton vehicles and smaller equipment would go by air with the troops, along with their individual combat gear, radios, and small arms. Selected larger pieces of equipment were approved for air movement on an individual basis. All other equipment was to move by sea. The aerial ports of departure were Camp Evans, Quang Tri, and Phu Bai.

On 29 October the Division completed and published an air movement plan which served as a guide as to when units would depart and the number of aircraft needed at origin and the destination of these aircraft. Due to the changes in the units stationing plan in III Corps Tactical Zone, it was necessary to make daily adjustments to this planned program. (The changes in the stationing plan were the result of later revised intelligence which caused General Forsythe to reorient some combat maneuver and aircraft assault units.) Even with these numerous changes, there was no case of an aircraft being on the ground available for loading without a load ready to be put aboard. As it turned out, aircraft loads were positioned well in advance of actual aircraft arrivals.

Units of the Division did not have definitive sea movement plans although they had made a sea movement into I Corps Tactical Zone just nine months earlier, utilizing eleven vessels. However, since that time, the Division had received a significant amount of additional equipment which made the previous sea movement experience invalid. The Division was offered initially more sea movement assets than it was capable of utilizing. Experience quickly indicated that the amount of regular incoming cargo, which was already back-logged on the ramps at Tan My, made it possible to pre-position only two LST loads without choking the port with outbound cargo. Communications with Tan My varied from poor to nonexistant. It was necessary for members of the Division G-4 staff to make several flights per day to Tan My to keep information updated and to properly move cargo forward based on the actual loading situation.

Container express containers were utilized to the utmost in packing cargo for sea movement. Units which had insufficient container express containers were provided pallets on which to load cargo. Incidentally, the Division had been urged for months by U. S. Army, Vietnam, to turn in a substantial number of on-hand container express containers. As it turned out, if these containers had been disposed of, the sea movement would not have been possible within the prescribed time frame.

Combat elements of the Division's 3d Brigade were the first to move and upon closing in the new area were placed under the operational control of the 1st Infantry Division, while combat elements of the 1st Brigade on closing were placed under the operational control of the 25th Infantry Division. The Division Headquarters deployed to Phuoc Vinh on 7 November and operational control of the 3d Brigade was returned to the 1st Cavalry Division. On 9 November, the Division assumed operational control of the 1st Brigade and combat elements of the 2d Brigade closed in the area under Division control on 12 November. The last combat essential equipment of the Division closed into III Corps Tactical Zone on 15 November 1968. Combat elements of the Division immediately began extensive reconnaissance in force operations throughout Tay Ninh, Binh Long, and Phuoc Long Provinces in an effort to stem enemy infiltration along the Cambodian border.

During Operation LIBERTY CANYON the 1st Cavalry Division used 437 C-130 sorties to move 11,550 passengers and 3,399 short tons of cargo. Additionally, a total of 31 LST's, three LPD's, three LSD's, and one aircraft carrier were used to transport 4,097 passengers and 16,593 short tons of cargo. The 1st Cavalry Division had moved on 24 hours notice over 570 miles by land, sea, and air, and immediately took over a large area of operations during a critical time. With the exception of thirty aircraft which were moved on the aircraft carrier, the remaining 415 aircraft flew all the way south. Many of them had been hastily repaired for a one-time flight.

However, the move was not without certain problems. Colonel William C. Dysinger, Commander of the Support Command for the Division, did not have enough movement control personnel to cover eight departure and arrival airfields and four seaports. However, he and Lieutenant Colonel Guinn E. Unger, the Division G–4, made up provisional teams that proved invaluable. The Saigon Support Command provided a liaison team at Phuoc Vinh with established special telephone circuits which became essential.

Theater support from all Services was timely and effective. General Irby was later to recall:

> The 1st Aviation Brigade did an outstanding job. They set up refueling, over-night stops, and food stations enroute. Our people could stop and get a rest break and something to eat and spare parts if they needed them . . . The Air Force and Navy were very responsive on the move. We coordinated on a minute-to-minute basis on the availability of C–130's, and the departure airfield control by the Air Force was outstanding.

General Forsythe summarized Operation LIBERTY CANYON as follows:

> From northern I CTZ to the jungle frontier areas of northwestern III CTZ, the Division moved and deployed to interdict the movement of four enemy divisions as they attempted to move souhward for an attack on the heartland of South Vietnam. The move was completed with great speed–(the first unit was in combat in III CTZ within 48 hours after it was alerted to move, and the Division was closed in 12 days)–and with strategic surprise (the enemy was confronted suddenly with a major force air assaulted astride his avenues of approach and into his prepared base areas with no prior warning). During the first nine days of the twelve day move, one brigade concurrently conducted and completed a major combat operation in northern I CTZ to penetrate the My Chanh Valley VC base area. In III CTZ the Division was given an immense area of operations (4,800 square miles); was based where existing space and facilities could be found (occupying 9 bases for our aircraft fleet); was given broad mission-type orders ("constitute the II FFV covering force"); and was given wide latitude and freedom of action to maneuver. In addition, care was taken not to fragment the Division or its assets, to permit the full range of its power, to find the enemy and to be shifted to fight him wherever he was found. It was proven beyond doubt that the total power of an airmobile division is greater than the sum of its parts. In short, we were given the opportunity to test the "theory of design and concept" of the airmobile division and we found it to be sound and practical.

CHAPTER XI

The Changing War and Cambodia, 1969-1970

The Changing War

Emphasis on the word "Vietnamization" after 1968 has tended to hide the fact that there was a great deal of mutual cooperation, training, and planning all along. In literally hundreds of specific areas, we had already long ago agreed that the Vietnamese should have full control of operations; and, in the Delta, which we will review in this chapter, it had always been the modus operandi. Notwithstanding, we saw the need to do much more. The U. S. Army increased existing programs to train the South Vietnamese in all aspects of airmobility including pilot and crew instruction, joint operations, in many cases from collocated U. S. and Army of the Republic of Vietnam division and brigade bases, and joint logistics. The operation into Cambodia in 1970 was an extremely successful example of what we could do together. The outstanding actions of all Army of the Republic of Vietnam units involved showed us that we were on solid ground in phasing our responsibilities over to the South Vietnamese. This was followed by a successful return to the same area the following year and by an incursion into Laos—LAMSON 719—in which Army of the Republic of Vietnam forces took on the very best that the North Vietnamese could muster and came out on top. This phase has been marked by the best spirit and mutual cooperation that the war has yet seen and by a dramatic tightening of the airmobile capabilities of Army of the Republic of Vietnam forces.

Supporting the Army of the Republic of Vietnam in the Delta

Earlier in this volume we touched on U.S. operations in the Delta region, but our examination of airmobility in Vietnam would be incomplete without mentioning the extraordinary accomplishments of our Army aviation units who for many years had been supporting the Army of the Republic of Vietnam operations in the IV Corps Tactical Zone. In several ways, their contributions were unique. Brigadier General George W. Putnam, Jr., who took command of the 1st Aviation Brigade on 6 January 1970 from Major General Allen M. Burdett, Jr., was to remark:

> . . . The real story of the Aviation Brigade is in the 164th Group in the Delta. Elsewhere it was, 'give so many helicopters here; and so many there.' The CG, 1st Aviation Brigade, exercised very little control over the assets of the Brigade in the I, II, and III Corps. But the 164th Group was not precisely controlled. Its commander could move assets; organize task forces . . . They had an organization for combat which permitted a diversity of aviation assets to support three ARVN divisions.

In December 1969 Colonel William J. Maddox, Jr.,[1] was assigned as Commanding Officer of the 164th Aviation Group, after commanding the 3d Brigade of the 25th Infantry Division. It was a fortuitous choice since Colonel Maddox had had extensive experience in the Delta—first, as Commanding Officer of the 13th Aviation Battalion, from July 1965 to August 1966; and, second, as Senior Advisor to the 21st Army of the Republic of Vietnam Infantry Division, from September 1966 to June 1967. Besides the unique geographic features, the big difference between the operations in the IV Corps Tactical Zone and the other areas of Vietnam was the lack of any long-term division-size U. S. troop commitment. This made the aviation group commander in a very large sense the "airmobile commander" in the Delta.

When Colonel Maddox returned to the Delta in 1969 he found that the allocation of the Group's 570 aircraft had not changed materially since the time he had been the Commanding Officer of the 13th Aviation Battalion, and he instituted major organizational changes to make airmobile support more responsive. The key change was the decentralization of the aviation assets into task forces—flexible organizations built around the assigned battalion and squadron headquarters. The 13th Aviation Battalion became Task Force GUARDIAN in support of the 21st Army of the Republic of Vietnam Division. Task Force COUGAR was built around the 214th Aviation Battalion and supported both the 9th and 7th Army of the Republic of Vietnam Divisions. The Army of the Republic of Vietnam forces in the border provinces, which had been desig-

[1] Colonel Maddox, one of the most decorated officers in the United States Army, was to be promoted to Brigadier General in July 1970 and become Director of Army Aviation on 18 September 1970.

nated the 44th Special Tactical Zone, were supported by the 7th of the 1st Cavalry, organized as Task Force BLACKHAWK. A general support task force was organized at Can Tho with the assets of the 307th Aviation Battalion. This latter unit was assigned the general support mission because it had the Chinook and Mohawk companies. This task force organization had a marked impact on the effectiveness of airmobile operations in the Delta and was warmly welcomed by the Army of the Republic of Vietnam commanders and their senior advisors. Colonel Maddox insisted that the battalion commanders visit each division they supported on a daily basis for personal coordination with either the division commander or his senior advisor.

The 13th Aviation Battalion had a long and proud history of airmobile support in the Delta even before there was a major U. S. presence in Vietnam. During the buildup in 1965, operations in the Delta received a lesser priority because of the overwhelming need to provide aviation assets to the major U. S. units arriving in country. Nevertheless, throughout this growth period the 13th Aviation Battalion was supporting major Army of the Republic of Vietnam operations against a well-organized powerful enemy.

In 1965 the IV Corps Tactical Zone was organized into 15 separate provinces. An additional province—Sa Dec—was formed in 1966. Each one of these provinces had an airfield at or near its provincial capital. Colonel Maddox set up a system of rearm and refuel points at each one of the fifteen airfields. This gave the 13th Aviation Battalion a "grid" across the Delta so that, with the exception of the extreme southern tip of the Ca Mau Penisula, no aircraft was more than twenty minutes away from a place where it could be rearmed and refueled.

Colonel Maddox would recall that when he arrived in 1965 refueling was accomplished by sending fuel trucks to the nearest airfield in the proposed area of operation on D-1. This was an obvious message to the Viet Cong that an operation was to take place and was a completely unreliable method of establishing a refueling point. Colonel Maddox pressured the appropriate Vietnamese and U.S. agencies to establish a pool of a million gallons of petroleum, oils, and lubricants which was broken out among the 15 airfields. Eventually, these same refueling points would handle rearmament as well.

In 1965 it was conservatively estimated that the Viet Cong had a strength in the Delta of 26 battalions including some very strong provisional battalions. One of these, known as the Tay Do Battalion in Can Tho Province, was considered to be a far more determined enemy than the Viet Cong main force units. During heavy contact with this enemy battalion in Operation Dan Chi 157, the 13th Aviation Battalion won the Presidential Unit Citation.

During 1965 and 1966 the three Army of the Republic of Vietnam divisions chipped away at the Viet Cong strength until they were able to reduce the size of their airmobile operations to company size units because of the lessening threat. The 21st Division, in particular, commanded by Colonel (later Major General) Nguyen Van Minh, was aggressive and imaginative in pursuing the hard core Viet Cong. Colonel Minh and Colonel Robert B. Spilman, who was the 21st Division Senior Advisor, were largely responsible for building the enviable record of the 21st Division.

The helicopter gunship became the one weapons system that could respond to the farflung outposts in a few moments notice day or night. The Vietnamese depended very heavily on gunship support and to a lesser extent on tube artillery. Operations in the Delta usually utilized three 5-helicopter gun platoons in the assault. One platoon would prepare a landing zone; a second platoon would accompany the lift ships into the landing zone; and the third platoon came in behind the lift ships to take over responsibility for fire support. Two officers of the 13th Aviation Battalion—Major Robert F. Molinelli and Captain Robin K. Miller—were singled out for official and unofficial recognition. Both of these individuals had made major contributions to gunship tactics in the Delta.

Major Molinelli and Captain Miller would be honored by being selected at different times by the Army Aviation Association as "Aviator of the Year." Space will not permit a full account of their accomplishments nor even a mention of the many other splendid young men who served in the Delta—especially, those brave soldiers who manned the lonely rearm and refuel points for months on end. However, they formed the backbone of the major airmobility innovation in the Delta—the dedicated task force concept.

Colonel Maddox, in explaining his rationale for forming task forces, remarked:

> . . . I have never agreed with blade time allocations at the lower levels of command; I have never believed in maintaining release times; I have never believed that an individual who supports should be overly concerned with the specific command relationship he has with the supported force . . . I found under the task force concept that the supported division actually became a proponent for his aviation support. Senior advisors and division commanders bragged to me that they had released assets early so that the aircraft could go home for daylight maintenance. I found in general that the relationships became very close and the level of support became much higher . . . We found

that we lost none of the flexibility that is necessary in shuttling aviation assets about the battlefield to meet tactical needs.

The Cavalry's Cambodian Campaign

Probably no single operation better demonstrated the airmobile concept than the 1st Cavalry Division's Cambodian campaign. Complete documentation of this important battle, to include the major Army of the Republic of Vietnam operation in the "Parrot's Beak," is an appropriate subject for a separate monograph. Here we will only be able to highlight this operation to bring out those salient points of the airmobility concept not touched upon in earlier chapters. This was the first example of a large-scale U. S. airmobile force in operation outside the borders of South Vietnam. It was the first time our commanders were allowed to cross the frustratingly close borders into the heart of the enemy sanctuary.

Beginning in the fall of 1968 the 1st Cavalry Division had straddled the enemy trails leading southward from the Cambodian border toward Saigon. The Viet Cong and North Vietnamese Army made desperate attempts to reestablish their logistical net in this area, with an obvious aim of repeating the attacks of *Tet* 1968. Beginning early in 1969 the 1st Cavalry fought a series of heavy skirmishes along these trails as three separate North Vietnamese Army divisions attempted to gain positions closer to the capital. The enemy effort was not successful. The 1st Cavalry's interdiction of the planned enemy operations for *Tet* 1969 is an untold story that merits much further study; although the enemy force was equally strong as it had been a year earlier during the infamous *Tet* attacks, the North Vietnamese Army plans were frustrated by the wide-ranging air cavalry surveillance and the superior mobility of the 1st Cavalry.

The enemy attempt during the early months of 1970 was weaker than the previous year. Nevertheless, several battalions did try to operate in force along the trail systems. At this time the 1st Cavalry area of operation covered 14,000 square kilometers. Airmobile troops kept a careful eye on five major north-to-south trails stretching from Tay Ninh Province across Binh Long and into the western two-thirds of Phuoc Long. There were several instances in which the cavalrymen uncovered base camps and fair-sized caches; however, it was evident that there was more to be found. Logistical trails of this size had to be supplied by warehouse type cache sites far more extensive than anything yet discovered.

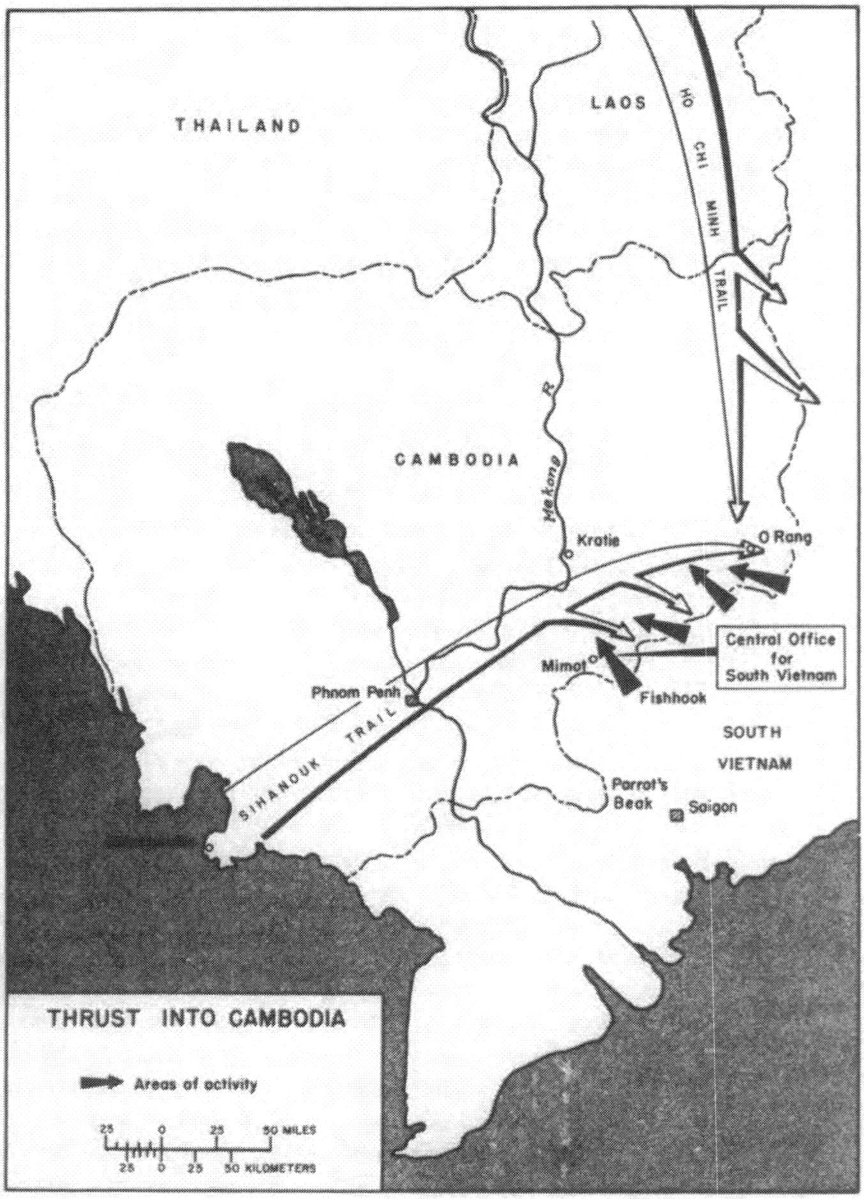

MAP 9

Although very little intelligence information was available concerning enemy operations in Cambodia, it was quite clear that a major enemy logistics effort was going on there. The staff of the 1st Cavalry felt that with the trails cut off the Division was causing the enemy to backlog supplies in Cambodia, increasing their already large depots. It was apparent that there could be no better time to strike at the heart of the enemy's supply base.

This area of Cambodia that borders III Corps Tactical Zone of the Republic of Vietnam had been used extensively by the Viet Cong and North Vietnamese since 1961 after the organization of Central Office South Vietnam, the headquarters commanding all Viet Cong and North Vietnam Army forces in III and IV Corps. During this period of nine years, the enemy had developed a series of base camps to house the administrative units of the Central Office South Vietnam headquarters plus a large number of associated units and subordinate commands. This included an artillery command, a security guard regiment, the 5th Viet Cong Division, the 7th North Vietnam Army Division, the 9th Viet Cong Division, and many service regiments. Although Allied forces had approached the border in large operations such as JUNCTION CITY, ATTLEBORO, and more recently the operations of the 1st Cavalry Division, the enemy had always had the advantage of being able to withdraw to the safety of his Cambodian sanctuaries.

The "Fishhook" of Cambodia extended from the generally flat plains adjacent to Mimot (*Map 9*) northeast through roughly rolling plains; and east to the roughly dissected hills and low mountains near O'Rang. Multi-canopied, dense undergrowth forest was the dominant natural vegetation throughout the area. Rubber plantations were found primarily in the western section. Open areas with dry crops, upland rice or marshes were scattered throughout. This area, because of its inaccessibility and the dense vegetation cover was extremely favorable to guerrilla warfare and restricted heliborne and mechanized operations. Generally, concealment from both aerial and ground observations was excellent.

The sequence of events which unfolded during the weeks which preceded the operation—the removal of Prince Norodom Sihanouk as Cambodian Chief of State, the closing of the port of Sihanoukville, and limited allied operations in other portions of Cambodia—had alerted the enemy to possible cross-border operations. Nevertheless, the enemy was not prepared for the massive assault to follow, nor was he prepared for an assault in depth. It is clear now that the North Vietnam Army felt that any crossing of the border would be a shallow operation of perhaps two or three kilometers. He was to discover, to his consternation, that the allied forces were at times twenty kilometers to his rear.

On 26 April 1970 Major General Elvy B. Roberts, the Division Commander, received instructions from Lieutenant General Michael S. Davison, Commanding General, II Field Force Vietnam, to prepare plans for a coordinated attack to neutralize the Central Office South Vietnam base area in the "Fishhook" of Cambodia. He was told that the 1st Cavalry should be prepared to implement this operation within 72 hours of notification.

This operation, here referred to as the "Cambodian Campaign," was officially entitled "Operation TOAN THANG 43, TOAN THANG 45, and TOAN THANG 46." It did not include the Army of the Republic of Vietnam operations in the area of the "Parrot's Beak" to the south which were concurrent and continued after the operations of the 1st Cavalry Division.

During the period 26 to 28 April, the 1st Cavalry Division and the Army of the Republic of Vietnam Airborne Division conducted joint planning for the operation. The major consideration at that time was the allocation of sufficient forces to insure successful accomplishment of the mission while continuing to conduct tactical operations within the III Corps Tactical Zone. The Allied forces that were to be used for the Cambodian Campaign were then deployed against the northern tier of III Corps Tactical Zone. The 1st Brigade of the 1st Cavalry was deployed in the far western War Zone "C"; the 3d Army of the Republic of Vietnam Airborne Brigade was in central War Zone "C"; and the 11th Armored Cavalry Regiment and 3d Brigade, 1st Cavalry, were in the eastern portion. The 9th Regiment, 5th Army of the Republic of Vietnam Division was operating in Binh Long Province with the 2d Army of the Republic of Vietnam Airborne Brigade to their east. The 2d Brigade of the Cavalry was in Phuoc Long Province. The Division Artillery Commander was responsible for the defense of a large sector centering on the division headquarters at Camp Gorvad.

Essentially, an attack into Cambodia meant little change in the operations of the 1st Cavalry. The Division had been moving progressively up to the border and expanding its interdiction operations both to the east and the west. Small hasty fire bases, each established only for a few days, had become the method of operation. Company and platoon-size airmobile units fanned out through wide areas of jungle and forest, travelling light, receiving resupply only once every three days. Since the division was already concentrating on fast-moving, light operations, leap-frogging from one small hasty fire base to the next, the order for the Cambodian campaign simply told it to do more of the same.

On 28 April, the division was further directed to be prepared to commence operations within 48 hours of notification. It had been decided that a combined task force would make the initial assaults into Cambodia. Command and control of this operation was given to the Assistant Division Commander for Maneuver of the 1st Cavalry Division, Brigadier General Robert M. Shoemaker. A combined U.S. and Army of the Republic of Vietnam staff was assembled at this time and prepared the final plans for the operation.

For Bob Shoemaker, this would be the culmination of years of dedicated effort to prove the airmobility concept. Ever since the early days of the 11th Air Assault Division and the movement of the 1st Cavalry to Vietnam, he had been recognized as one of the foremost tacticians of airmobility.

The concept of the operation was that Task Force Shoemaker, consisting of the 3d Brigade with one mechanized infantry battalion and one tank battalion under operational control, the 3d Army of the Republic of Vietnam Airborne Brigade, and the 11th Armored Cavalry Regiment, would conduct air assaults and ground attacks into the "Fishhook" of Cambodia. Following an intensive preparation phase of B–52 strikes, tactical air strikes, and artillery bombardment, the 3d Army of the Republic of Vietnam Airborne Brigade would air assault into the area north of the objective to seal off escape routes and begin operations to the south. Simultaneously, the task force (–) would attack north across the Cambodian Border with the 3d Brigade on the west and the 11th Armored Cavalry Regiment on the south and east. The 1st Squadron, 9th Cavalry (–) would conduct screening operations in the task force area of operations while elements of the 9th Army of the Republic of Vietnam Regiment and the 1st Army of the Republic of Vietnam Armored Cavalry Regiment would screen to the east in Binh Long Province adjacent to the objective area. All elements would then conduct search and interdiction operations to locate and exploit enemy lines of communication and cache sites in the objective area.

In the early hours of 1 May, six serials of B–52's dropped their heavy ordnance on hard targets within the primary objective area. The last bomb went off at 0545. Fifteen minutes later an intense artillery preparation began with the priority to the proposed landing zones in the 3d Army of the Republic of Vietnam Airborne Brigade's objective area. D-day had arrived.

At 0630 the 1st Army of the Republic of Vietnam Cavalry Regiment began its movement from the northwest of An Loc toward the border. At the same time a *15,000 pound* bomb, with an extended fuse designed to detonate about seven feet above the ground, was dropped to clear the jungle at landing zone EAST. This was followed fifteen minutes later by a similar drop at landing zone CENTER. Shortly after first light, the Forward Air Controllers began directing tactical air strikes on pre-planned targets, shifting to the 3d Army of the Republic of Vietnam Airborne Brigade's objective area during the period from 0700 to 0800.

The 1st Squadron, 9th Air Cavalry began aerial reconnaissance operations early on D-day and by 0740 had established contact. Five North Vietnam Army soldiers and their 2½-ton truck became the first recorded casualties of the operation. At 0800 hours the 1st Squadron, 9th Cavalry conducted a landing zone reconnaissance which was followed ten minutes later by the combat assault of an Army of the Republic of Vietnam airborne battalion into landing zone EAST. The landing zone was secured and became a fire support base when six 105-mm howitzers and three 155-mm howitzers were inserted shortly thereafter. During this air assault, the 11th Armored Cavalry Regiment had moved out of their staging area and crossed the line of departure, moving north. In the 3d Brigade area, C Company, 2d Battalion, 47th Infantry (Mechanized) crossed the Cambodian border at 0945, followed by elements of the 11th Armored Cavalry Regiment to the east which crossed approximately fifteen minutes later.

The 5th Army of the Republic of Vietnam Airborne Battalion in a 42-ship lift supported by 22 Cobra gunships began its combat assault into Objective "B" at 0946 and reached landing zone CENTER by 1005. The 9th Army of the Republic of Vietnam Airborne Battalion completed its combat assault into Objective "A" on the west.

Contact was immediately established with a panicked North Vietnam Army force of approximately 200 men. The Cobras supporting the contact expended most of their rockets and machine gun ammunition on groups of 10 to 30 North Vietnam Army men fleeing the area in a dozen directions. It was apparent that tactical surprise had been achieved during the combat assaults as there were no reported instances of .51-caliber ground-to-air firing.

During the afternoon of D-day, two companies of the 2d Battalion, 7th Cavalry made a combat assault into Objective X-RAY in the northern portion of the 3d Brigade area of operation. This movement had been tentatively planned by General Shoemaker and, due to the relatively light resistance throughout the area, he ordered its execution as the final combat assault of D-day.

The enemy reaction to the opening of the Allied Offensive took the form of a confused, milling crowd, ill-prepared to deal with the massive onslaught that was unleashed. Tactical surprise was complete. The enemy had not left the area, nor had he reinforced or prepared his defenses. The heliborne assault forces were not greeted with heavy anti-aircraft fire but rather only with small arms fire from a few individuals. Nowhere in evidence were the heavy machineguns from the three antiaircraft battalion-size units known to be in the area. While later evidence showed that while some strategic preparations had been made hedging against a possible allied thrust, the enemy tacticians had not taken steps to counter an air assault. Airmobility had again caught the enemy off-balance. The results were evident, as noted in the following official excerpts of the day's activities:

> The 1/9 Cav had a field day catching small groups of NVA trying to evade, resulting in a record total of 157 NVA killed by helicopter.
>
> TAC Air in another record setting day put a total of 185 sorties on hard targets which resulted in 109 NVA KBA in the ARVN Airborne AO alone.
>
> Among the ARVN Airborne forces, the 5th Battalion was outstanding with 27 NVA killed and 8 prisoners taken during the day. The prisoners were later identified as members of the 250th Convalescence Battalion, the 50th Rear Service Group and the 1st Battalion, 165th Regiment, 7th NVA Division.
>
> The 3d Company, 3d ARVN Airborne made the first significant cache discovery at 1720 hours when they found a large medical cache of up to 6,000 pounds. The cache included the finest in modern surgical equipment and had been imported from western Europe via Air France, possibly through Phnom Penh.
>
> The ground contact of Company H of the 2/11 ACR was the highlight of the 11th ACR operations during the day. After passing through a regimental-size base camp, a large enemy force was encountered in trenches to the north. The ensuing battle left 50 enemy dead versus 2 U. S. KIA, the only U. S. combat fatalities of D-day.

The next few days of operations were characterized by a continuation of maneuvers begun on D-day. The enemy made strenuous efforts to avoid contact and to determine the extent and placement of the Allied forces. His command and control apparatus was completely disrupted and he was caught off guard and ill prepared. The High Command scattered in two's and three's and a large exodus of trucks going in all directions was noted by the 1st Squadron, 9th Cavalry.

On 3 May Task Force Shoemaker was reinforced with elements of the 2d Brigade. Multiple small caches were being discovered by the ground units while the first large weapons cache was observed from the air by A Troop of the 1st Squadron, 9th Cavalry. This area was engaged with gunships and Tactical Air resulting in the

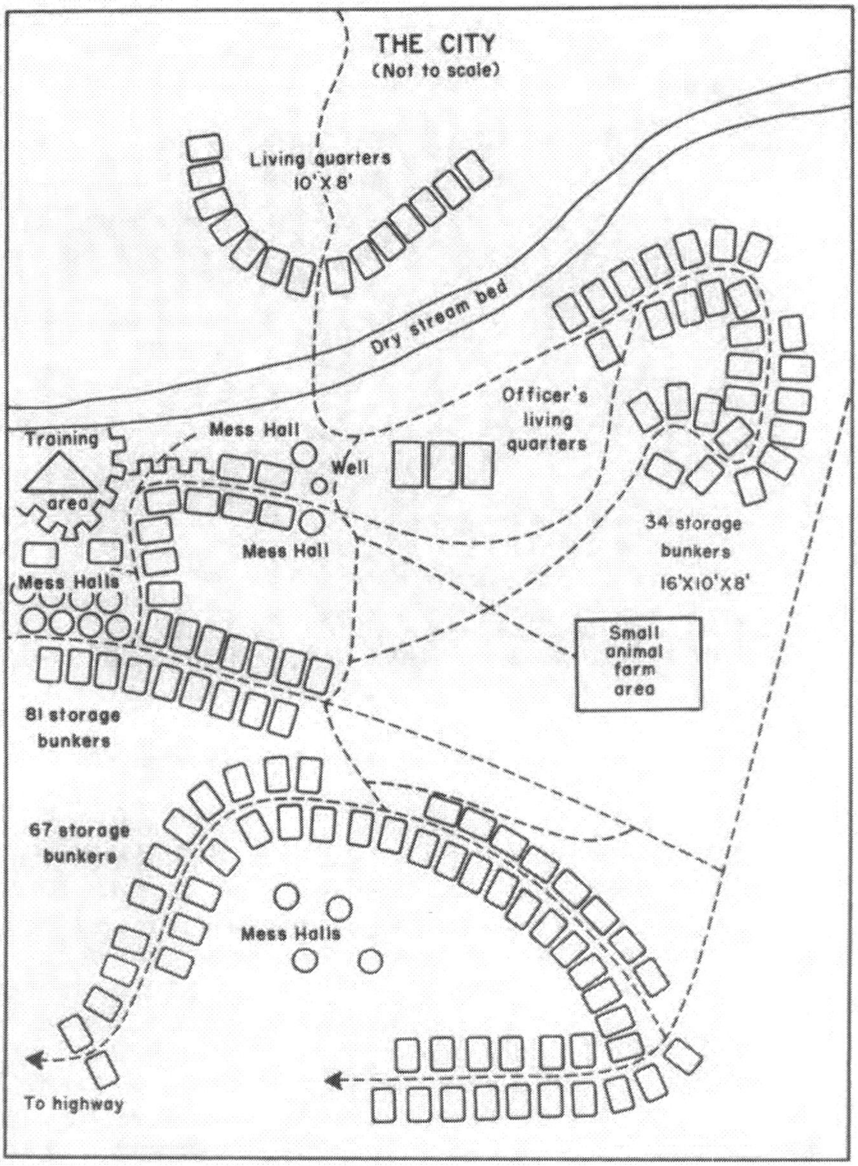

MAP 10

destruction of 7 2½-ton trucks, 13 ¾-ton trucks, and 3 jeeps. Another truck park in the nearby area was discovered and the Cavalry Troop destroyed nine trucks with their own gunships.

On 4 May 1970, B Troop of the Cavalry Squadron observed numerous bunkers and military structures in a densely vegetated

CAMBODIANS FILL BAGS WITH CAPTURED RICE, 18 MAY 1970

area northwest of the current ground operations. Additional aerial reconnaissance teams further reported that these structures and bunkers were connected with bamboo matted trails. One pilot also reported seeing numerous antennas in the southern part of the complex. On 5 May, Company C, 1st Battalion, 5th Cavalry, in response to the aerial reconnaissance sighting reports, displaced into the northern half of the complex which had been dubbed "the city". Immediately upon entering the suspected area, the Cavalry troopers reported finding numerous storage bunkers measuring 16 feet long, 10 feet wide, and 8 feet deep. Subsequent reports indicated that most of the bunkers contained large quantities of weapons and munitions and that the enemy had just recently evacuated the area. Throughout the period 5–13 May 182 storage bunkers, 18 mess halls, a training area and a small animal farm were discovered in the area.

The logistical storage bunkers contained clothing, food stocks, medical supplies, weapons and munitions. The bulk consisted of ammunition. Generally, all types of equipment and supplies were in an excellent state of preservation and in good operating condition when discovered. "The city" (*Map 10*) covered approximately three square kilometers and consisted of a well-organized storage depot that was capable of rapid receipt and issue of large quantities of supplies. Judging from the general condition of the oldest bunker and from translation of supply documents found in the area, it was apparent that the storage depot had been in operation for two to two and one-half years. Some bunkers had been constructed within the last six months.

Captured supply records indicated that the supply depot primarily supported the 7th North Vietnam Army Division. Based on the discovery of classroom facilities, numerous mess halls, firing ranges, as well as large stocks of items of personal clothing and equipment in the cache site area, it was determined that "the city" was also used to provide refresher military and political training to recent replacements from North Vietnam.

Selected items of equipment captured at this cache site complex included:

1,282 individual weapons
202 crew-served weapons
319,000 rounds of .51-caliber/12.7-mm ammunition
25,200 rounds of 14.5-mm antiaircraft machinegun ammunition
1,559,000 rounds of AK–47 ammunition
2,110 grenades
58,000 pounds of plastic explosives
400,000 rounds of .30-caliber ammunition
16,920 propelling charges for 120-mm mortar
22 cases of anti-personnel mines
30 tons of rice
8 tons of corn
1,100 pounds of salt

Throughout the Cambodian campaign, Allied forces would uncover other major caches of equipment which proved that this area was truly one of the most important logistical bases of the enemy. On 25 May a large automotive parts cache was discovered by A Company of the 2d Battalion, 8th Cavalry which was the first of a series of caches of various supplies discovered in the area from 25 May to 9 June. One of these, a communications depot, discovered by D Company, 2d Battalion, 8th Cavalry, indicated that the enemy did not have time to evacuate his most valuable communications equipment. Considering the critical nature of enemy signal

A CH–47 CHINOOK HELICOPTER LIFTS OFF A SLINGLOAD OF AMMUNITION FROM FIRE SUPPORT BASE MYRON IN CAMBODIA, 24 JUNE 1970

equipment, it was felt that his equipment would have received top priority for evacuation. Like "the city" cache site, this area also had supplies segregated, by type, to facilitate storage and distribution to receiving units. Equipment and supplies were for the most part new and in excellent condition.

Unlike the Ho Chi Minh Trail, the Cambodian caches were filled with material transported for the most part by truck. Truck repair centers and spare parts were part of the logistics complex. The 1st Cavalry had deprived the enemy of over 305 vehicles in its Cambodian operations. While a few of these were primarily for passengers, such as a captured Porche, Mercedes-Benz and jeeps, the vast majority were cargo carriers. These trucks had a total capacity of 442 tons. Intelligence showed that the captured vehicles were only a small part of the North Vietnam Army truck inventory in Cambodia. During the early days of the operation the Air Cavalry Squadron reported that many Cambodian roads showed heavy use of trucks away from the area of operations. Obviously, they had been among the first items to be moved out of the area.

The size of the North Vietnamese Army logistics system in Cambodia adjoining III Corps Tactical Zone was vast. It had the capacity to move thousands of tons of material from various points in Cambodia to supply depots along the South Vietnamese border and move these supplies quickly over the network of roads that connected the various caches. The system was so dependent on trucks that an extensive refueling and repair organization was necessary. Sophisticated and effective, the North Vietnam Army supply system was a major weapon in the enemy's arsenal.

By mid-May, the search had expanded eastward to the border area north of Phuoc Long Province where on 8 May the 2d Brigade discovered a new significant cache site. This North Vietnam Army base camp, nick-named ROCK ISLAND EAST, eventually yielded 329 tons of munitions. As the operation continued, the 1st Brigade was moved from War Zone C to the O'Rang area east of the 2d Brigade. Both Brigades used the airstrip at Bu Gia Map as a forward logistics base. One rifle company patrolled out of the abandoned Special Forces camp there as security for the refuel, rearm point and the tons of palletized supplies which were offloaded from C-130's and C-123's.

The withdrawal of the 1st Cavalry from Cambodia, although the most critical of all the operations, was executed in a truly classic manner. The withdrawal sequence was time-phased to allow for the redeployment of one fire support base each day. This phasing would allow for even utilization of aircraft assets, particularly the CH-54 Cranes of the 273d Aviation Company which were required for the movement of bridges, 155-mm howitzers, 2 ½-ton trucks, and bulldozers. On the second day of extraction, while lifting the 1st Battalion, 50th Cavalry from Fire Support Base DAVID, the aviation units ran into extremely poor weather with ceilings at zero, fog and rain. The Chinooks from the 228th Assault Support Helicopter Battalion showed the ultimate in professionalism by flying at extremely low levels through the valleys, skirting the fog-covered hills, and extracting the unit. On 24 June, while extracting Fire Support Base BRONCHO, one aircraft was downed and five others were hit. At the same time the fire support base was taking indirect fire regularly throughout the day. Using all possible suppressive fires from the armed helicopters, Fire Support Base BRONCHO was extracted at last light.

Besides the Division's organic Cranes, the 273d Aviation Company (Heavy Helicopter) was under operational control of the 1st Cavalry during the sixty days of the Cambodian Campaign. This company with its CH-54 Flying Cranes lifted essential engi-

Troops Descending an Aerial Ladder Into Triple Canopied Jungle

CH–47 Chinook Delivering Captured Rice in Republic of Vietnam Controlled Rural Area

neering equipment (272 bulldozers, 54 backhoes, and 41 road graders) as well as all 155-mm howitzers into (and out of) the operational area. They moved bridge sections and recovered $7,315,000.00 worth of downed aircraft. During the Cambodian Campaign, this company flew 2,486 sorties, or, as a logistician might put it, 177,688 ton-miles.

On the final day of the operation, in the actual crossing of the border by all U. S. troops, every possible precaution was taken to insure success. Troop ladders, smoke ships, pathfinders, and recovery aircraft were available to cover any contingency. The crossing proved uneventful with the last CH–47 aircraft leaving Cambodia at 1523, 29 June. The honor of being the last U. S. Army aircraft out of Cambodia went to B Company of the 1st Squadron, 9th Cavalry whose screening "Pink Team" reported reentering Vietnam at 1728, 29 June.

The 1st Squadron, 9th Air Cavalry proved again during the Cambodian Campaign how invaluable this capability is to any airmobile operation. During the period 1 May 1970 to 30 June 1970 the Squadron had performed intensive ground and aerial reconnaissance operations almost every flyable hour. The Squadron's assets were shifted as necessary, capitalizing on mobility, reconnaissance, and firepower in order to determine enemy locations and escape routes. Using "Pink Teams"—one Cobra gunship and one OH–6A observation helicopter—the Air Cavalry troops were able to cover large areas effectively. When the situation warranted, the aero-rifle platoon would be inserted to face the enemy until a larger force could be committed into the area. The intelligence provided by the 1st Squadron, 9th Cavalry enabled the division to redeploy its assets and effectively destroy many of the enemy's large cache sites.

The 1st Cavalry Division was not sure that they would be able to evacuate or destroy all the large quantities of enemy supplies that were found during this operation; but, as the engineers repaired roads and constructed bridges to allow a convoy evacuation, these problems were solved. The enormous quantities of rice discovered were distributed to the various agencies under the watchful eyes of the G–5. This captured rice was also used to feed the large number of refugees that came pouring into the Cavalry area of operations.

The operation had revealed several previously unknown facts and confirmed several suspicions. The size and locations of the enemy's base areas were plotted and for the first time the extent and depth of the enemy's logistical system was documented. Whereas the base areas had previously been thought to all be close to the border, it was discovered that the depth extended far into Cambodia, often serviced by unknown roads and trails built specifically for this purpose. The existence of recovery and replacement regiments as well as the organization and operations area were discovered. Vast quantities of new weapons and munitions were captured, possibly preempting their intended use against Allied forces. During the Cambodian Campaign, the 1st Cavalry Division (including those units under its operational control) accounted for 2,574 enemy killed in action and 31 prisoners of war. They captured 2,244 tons of rice and over ten million rounds of ammunition.

The 1st Cavalry Division operation in Cambodia far exceeded all expectations and proved to be one of the most successful operations in the history of the First Team. All aspects of ground and air combat were utilized—air cavalry, armor, infantry, and mechanized infantry. The U.S. Air Force reconnaissance, tactical air, and B–52's performed yeoman duty throughout the campaign there. This team effort, spearheaded by the airmobile flexibility inherent in the 1st Cavalry, carried the war to the enemy and defeated him in his own backyard.

Major General George W. Casey had taken command of the 1st Cavalry on 12 May from General Roberts. General Roberts, who had been scheduled to turn over command of the division on 3 May, had been extended until 12 May in order to plan, organize, and command the initial phases of the Cambodian operation. On 6 July 1970, he wrote a letter to his troops summing up the 1st Cavalry's part in the Cambodian Campaign. Excerpts from the letter are as follows:

> . . . The results are impressive. You killed enough of the enemy to man three NVA Regiments; captured or destroyed enough individual and crew-served weapons to equip two NVA Divisions; and denied the enemy an entire year's supply of rice for all of his maneuver battalions in our AO. You captured more rocket, mortar, and recoilless rifle rounds than the enemy fired in all of III Corps during the twelve months preceding our move into Cambodia. And, perhaps most important, by working together in an airmobile team, you disrupted the enemy's entire supply system, making chaos of his base areas and killing or driving off his rear service personnel.
>
> Only time will tell how long it will take the NVA to recover, but of this you can be sure—you have set the enemy back sufficiently to permit President Nixon's redeployment plan to proceed with safety while assuring that our Vietnamese Allies maintain their freedom. This is your achievement. This is yet another demonstration that you of the 1st Cavalry Division deserve—and have earned again—the accolade of the FIRST TEAM. It is my honor to have served alongside you during this crucial and historic period.

The following day on the morning of 7 July, General Casey was enroute to Cam Ranh Bay to visit wounded Sky Troopers in the hospital. Flying over the rugged mountains of the Central Highlands, General Casey's helicopter entered a thick cloudbank and disappeared from sight. In the late afternoon of 9 July, the wreckage of the General's helicopter was found. General Casey and all the officers and men aboard had been killed instantly in the crash. George had served with me as my Chief of Staff during 1967. Prior to that he commanded the 2d Brigade under General Norton. His death was a great personal loss to me and everyone who had served with him. Major General George Putnam, whose 1st Aviation Brigade had been doing a fantastic job in supporting Army of the Republic of Vietnam operations in the Parrot's Beak to the south, was ordered to take command of the 1st Cavalry Division.

CHAPTER XII

Organizational Changes and Laos, 1970-1971

Organizational Changes

With the exception of decentralizing its maintenance, the organization of the 1st Cavalry Division had remained essentially unchanged since its deployment to Vietnam. Indeed, General Howze would find its organization very similar to the proposed plan his Board had prepared in 1962. However, after the Cambodian campaign, the ever-increasing area of operation, and the requirement to support more Army of the Republic of Vietnam operations, General Putnam was prompted to examine means to increase his air cavalry capability.

In August 1970, General Putnam directed an analysis of the productivity of the aircraft assets of the 1st Cavalry Division. This analysis disclosed that airlift escort by a section of two Cobras from the gun company of the airlift battalions was the least productive mission being flown by the division. It was determined that the escort at that time could be forgone since aerial rocket artillery ships were always at the critical points, the pickup zones and landing zones. The analysis also revealed that essential general support missions normally flown by the OH-6A could be supported by fewer aircraft if careful controls were maintained. Based on these findings, two provisional air cavalry troops were formed using the Assault Weapons Companies of the 227th and 229th Assault Helicopter Battalions and attaching necessary OH-6's and personnel from other Division units. This enlarged the air cavalry squadron to five troops and greatly increased the Division's capability to cover its farflung operations.

A short time later, the 1st Cavalry Division was given operational control of a separate air cavalry squadron, the 3d Battalion 17th Cavalry. General Putnam commented:

> I then had two and two-thirds squadrons of air cavalry. Our ARA battalion had always been responsive to fire support requirements from

the air cav. But when we began supporting ARVN divisions with air cav (and the air cav under my OPCON), I then gave the ARA battalion the additional mission of supporting this air cav. What I had in essence was an Air Cavalry Combat Brigade as originally conceived by the Howze Board. This proved to be tremendously successful in supporting the ARVN in Cambodia.

During this same period in late 1970 the 1st Cavalry Division introduced new airmobile tactics in using the 81-mm mortar. The 81-mm mortar, long a valuable weapon to the Infantry, was used by the 1st Cavalry to support strike operations outside of tube artillery range. The mortar, which required a smaller security element than an artillery base and could be supported by the Huey, was established in a temporary mini-base located on the periphery of regular artillery range to extend indirect fire in support of ground troops. This became increasingly important as the number of squad and platoon-size operations increased.

After the Cambodian Campaign, it became the rule rather than the exception to conduct small unit operations down to separate squad and platoon-size forces, rather than the multi-battalion operations of previous years. In this way the Cavalry could cover a larger area more thoroughly, but this method of operation brought with it the requirement for a high caliber of leadership at the lowest level. The young Cavalry lieutenants and sergeants more than adequately proved they were up to the job. These small unit operations were enhanced by the inherent capability of the division to reinforce rapidly and the great flexibility and variety of firepower at its disposal.

As an example of the firepower available at this time, the standard armament of the Cobra now included the 2.75-inch rocket with a 17 pound warhead, the very effective 2.75-inch Flechette rocket, and the SX–35 20-mm cannon. The firepower of the division was enhanced by the intelligence gathering capability of the Seismic Intrusion Devices which were dropped by UH–1H helicopters along known infiltration routes. Once enemy movement had been detected, a small unit was lifted into an area well ahead of the enemy's determined course of movement and established an effective ambush with artillery and gunships standing by.

Into Laos

The final airmobile operation to be included in this study was given the code name of LAMSON 719. This combined operation took place in Laos from 8 February to 9 April 1971. LAMSON 719 was unique in many ways, but of principal concern to this study was the impression, generated both in and out of the military by the early reports of severe helicopter losses, that the airmobile concept had "fallen flat on its face"—that airmobility brought unacceptable risks when subject to any threat more than low-intensity antiaircraft fire in the "permissive" environment of South Vietnam. As is so often the case, the impact of the initial headlines remained uncorrected by the later objective review of the facts. Many believed that this operation was nothing short of a disaster when, in fact, it proved again the basic soundness of the airmobile concept and scored a devastating blow to the enemy's logistics sanctuary in Laos.

In the next few pages I've made no attempt to relate the full story of LAMSON 719—the detailed ground battle between the communist forces and the Army of the Republic of Vietnam. That is properly a Vietnamese story—to be recounted elsewhere. For the record, the Army of the Republic of Vietnam fought tenaciously against ever-increasing odds and reached their objective. The Laos operation was a tactical and strategic success, as well as a psychological success, for the Republic of Vietnam.

Before one draws any comparisons between the Laos operations and airmobile operations conducted by the U. S. Army, it must be realized that LAMSON 719 was a very special operation in which strict rules governed U. S. military operations across the Laotian border. While the Republic of Vietnam Armed Forces could operate freely on the ground and in the air within Laos, U. S. Forces were restricted to air operations under specific rules of engagement and were prohibited from fighting on the ground.

The fact that U. S. personnel were forbidden to go on the ground in Laos required modification of normal procedures for supporting firepower, coordination and conduct of airmobile operations, and rescue and recovery of downed crews and aircraft. The absence of U. S. advisors with the ground forces and the language difficulties added further complications.

Furthermore, LAMSON 719 was a combined operation in which Lieutenant General Hoaug Xuan Lam, the Commanding General, I Corps, Army of the Republic of Vietnam, planned and conducted the ground operations in Laos while Lieutenant General James W. Sutherland, the Commanding General, XXIV Corps, U. S. Army, planned, co-ordinated, and conducted airmobile and aviation operations in support of Republic of Vietnam Armed Forces ground operations. Though these two commanders developed a high order of co-operation and mutual confidence, there was an absence of the unity of command of ground and airmobile forces that characterized airmobile operations conducted unilaterally by the United States Army.

By late September and early October 1970 it had become obvious from various enemy actions and intelligence sources that the North Vietnamese Army planned to strangle Phnom Penh and overthrow the Lon Nol Government. At the same time, there was ample evidence that the North Vietnam Army would continue its aggression against South Vietnam and rebuild its bases along the Cambodian border adjacent to III and IV Corps Tactical Zones. The key to these enemy operations was an intensified resupply and reinforcement operation in southern Laos during the dry season which would last from mid-October to mid-April 1971.

December 1970 and January 1971 brought a sharp increase in the amount of supplies moved into the southern Laotian area known as Base Area 604, adjacent to Quang Tri Province in I Corps Tactical Zone. The intelligence community further noted that only a small portion of these supplies had been moved further to the south. In previous years the enemy had reached his peak efficiency in February and March in moving supplies down the Ho Chi Minh Trail. Accordingly, an attack against the base areas in Laos during these months presented the highest probability of inflicting the greatest damage to the enemy. Operation LAMSON 719 was conceived, developed, and implemented to react to this intelligence information.

Air interdiction of the entry points from North Vietnam into southern Laos had intensified since October 1970 and the 7th Air Force had been very effective in destroying enemy trucks. A new record of kills was reached in December and January. Army of the Republic of Vietnam operations into Cambodia were started in November 1970 with the mission of opening land and water routes to Phnom Penh. The Vietnamese forces had successfully expanded their area of operation and demonstrated their ability to conduct a major campaign without any advisory supervision.

In early February 1971 the Government of Vietnam decided to commit more than three Army of the Republic of Vietnam divisions to interdict the enemy's supply and infiltration routes in southern Laos and to destroy his logistical facilities and supplies. The broad objective was to reduce the North Vietnam Army capability for waging war in the south and to advance the security of the people of the Republic of Vietnam.

The operational area of LAMSON 719 covered an area roughly thirty-five to sixty kilometers. The geography of this area varied dramatically. The Xe Pon River split the area and was roughly paralleled by Highway Nine. Vegetation was mostly single or double canopy jungle along the river. Just south of the river rose a sheer escarpment leading to rugged mountainous terrain. Natural clearings were rare throughout the area and landing zones usually had to be carved out of the dense undergrowth. Intelligence indicated that the natural landing zones would be heavily defended.

The airmobile operations of LAMSON 719 were spread through three areas: the coastal base camps where most of the helicopters were kept at night; the forward staging area at Khe Sanh, where only a few helicopters remained overnight; and the operational area over Laos. Weather conditions at any one or all three locations could have a major effect on helicopter support. The right combination of weather conditions had to exist before helicopters could take off from the coastal bases, land at Khe Sanh to refuel and be briefed for missions, and fly into the operational area over Laos.

Early morning fog, rain and cloud cover, sometimes delayed airmobile operations until late morning or early afternoon. Rarely did weather conditions preclude operations all day throughout the operational area. On occasion, airmobile operations were conducted under ceilings and weather conditions that precluded employment of tactical air support. The smoke and dust raised by artillery combined with natural haze sharply reduced the visibility and frequently caused flying safety hazards. The highest degree of professionalism was required from all pilots.

The terrain features in the area, especially the higher elevation of the Annamite Mountain chain in the operational area, combined with the marginal weather to have a decided effect on airmobile operations. The river valleys, in particular the east-west oriented Xe Pon, became natural flight routes due to navigational requirements in marginal weather. This in turn focused enemy antiaircraft fire on obvious air routes.

The enemy forces in southern Laos were logistics organizations of Base Area 604, with reinforcements from regular North Vietnamese Army units. Besides the permanent service force of engineers, transportation, and antiaircraft troops, the North Vietnam Army forces included elements of five divisions, twelve infantry regiments, a tank regiment, an artillery regiment, and nineteen antiaircraft battalions. Each of the divisions had previously fought in South Vietnam and most of the enemy had taken part in the large-scale operations around Khe Sanh and Hue in 1967 and 1968. In summary, the enemy consisted of large conventional forces of

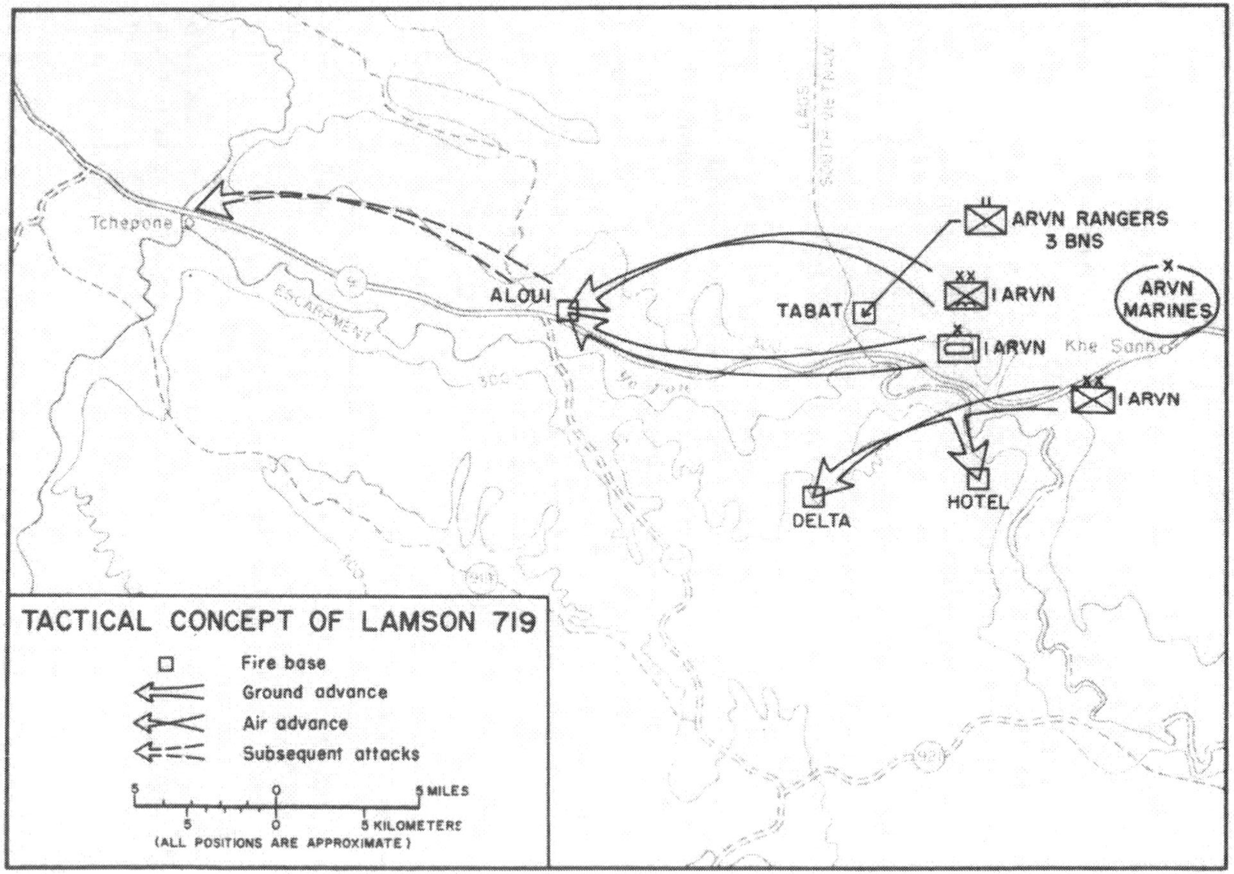

TACTICAL CONCEPT OF LAMSON 719
Fire base
Ground advance
Air advance
Subsequent attacks
5 MILES
5 KILOMETERS
(ALL POSITIONS ARE APPROXIMATE)
ARVN RANGERS 3 BNS
I ARVN
I ARVN
I ARVN
ARVN MARINES
Khe Sanh
TABAT
ALOUI
Tchepone
ESCARPMENT
DELTA
HOTEL
LAOS
SOUTH VIETNAM

MAP 11

infantry, tanks, and artillery capable of sustained mid-intensity conflict—lacking only air support.

The major Army of the Republic of Vietnam forces assigned to LAMSON 719 were the 1st Infantry Division, 1st Airborne Division, the Marine Division, three battalions of Rangers, and the 1st Armored Brigade with three cavalry squadrons. The U. S. elements operating in direct support of the Army of the Republic of Vietnam troops inside Laos consisted of the 2d Squadron, 17th Cavalry with four Air Cavalry troops, the 101st Aviation Group, with a number of aviation units under their operational control from the 1st Aviation Brigade, and one squadron of Marine medium transport helicopters.

The tactical concept for LAMSON 719 envisioned the Airborne Division, with the 1st Armored Brigade attached, making the main attack by air assault and overland movement astride Highway Nine to Aloui, and then proceeding in subsequent attacks to Tchepone. (*Map 11*) Highway Nine was to be opened as the main supply route. The 1st Infantry Division, according to the concept, was to attack on a parallel axis to the main attack along the high ground south of the Xe Pon River and protect the left flank of the Airborne Division. The Ranger group would establish a fire base, near the Laotian border north of Tabat, and protect the right flank of the Airborne Division. A Marine brigade was to be the reserve in the vicinity of Khe Sanh.

As early as January 1971, a planning group consisting of key staff officers from the U.S. XXIV Corps and Army of the Republic of Vietnam I Corps was established at Da Nang. Information on the operation was tightly held and, in some instances, this restriction of information complicated the preparation for LAMSON 719, especially in those areas where a long lead time was necessary. Though control and coordination procedures were agreed upon, it would not be until three weeks after Vietnamese troops had crossed the Laotian border that a combined tactical command post at Khe Sanh became a reality.

The Battle

The attack into Laos was initiated on 8 February from bases established on the Khe Sanh Plain. The Army of the Republic of Vietnam 1st Armored Brigade Task Force crossed the border at 1000 and advanced nine kilometers to the west along Route Nine on the first day. Three battalions of the 3d Regiment, 1st Army of the Republic of Vietnam Infantry Division, air assaulted into landing zones south of Route Nine while two battalions of the 1st Army of the Republic of Vietnam Airborne Division air assaulted north of Route Nine. Some 105-mm howitzer batteries were air-landed in both areas on D-day.

On 9 February, all air moves were cancelled due to adverse weather; however, the armored task force was able to move two kilometers further to the west. On 10 February, the 1st Army of the Republic of Vietnam Airborne Division air assaulted a battalion into Objective Aloui and the armored task force linked up with this battalion at 1555. On the same day, the 1st Army of the Republic of Vietnam Division landed a battalion at landing zone DELTA and the initial objectives of LAMSON 719 had been seized.

After the attack on 8 February the enemy reacted violently to the allied offensive. He aggressively employed his weapons and troops already present in Southern Laos and he reinforced heavily his forces and committed a variety of weapons including tanks to the battle. Reinforcements came from North Vietnam, South Vietnam, and other parts of Laos.

By 19 February the Rangers in the north were receiving frequent attacks by medium artillery, sappers, and infantry and resistance was stiffening in the area of the 1st Army of the Republic of Vietnam Airborne Division. Resupply and medical evacuation became increasingly more difficult. When weather precluded the employment of tactical air, as it often did until noon, and emergency resupply and medical evacuation was urgently required, the availability of helicopter gunships became even more critical.

By 22 February attacks against Fire Bases 30 and 31 and the Ranger positions were becoming more frequent and more intense. Enemy mines, ambushes, and the severe lack of maneuver room combined to slow the movement of the armor columns and they were unable to reach the Rangers to relieve the pressure. Consequently, it was decided to extract the Rangers on 25 February to a less hostile area near the Republic of Vietnam border. However, by this time, enemy supply bases one and two kilometers square had been found and a major petroleum, oils, and lubricants pipeline had been found and cut by Air Cavalry gunships. Tons of ammunition and food stocks had been destroyed. Six hundred and eighty weapons had been captured.

On 25 February the enemy made a classic armor attack against Fire Base 31. They had moved their armor stealthily over concealed routes to final assault positions before being discovered. Then the tanks with supporting infantry launched a violent daylight attack against the fire base. The defenders, supported by U. S.

tactical air, threw back the first and second waves of the enemy attack; but, on the third wave, three Soviet-made T–34 tanks made it to the top of the Base and forced the withdrawal of the defenders. This was to be the first and last success of enemy tanks during LAMSON 719 and the only friendly fire base to be completely overrun in Laos.

Three Army of the Republic of Vietnam armored cavalry squadrons and four infantry battalions had not proved sufficient to provide ground security for the 20 kilometers of road in Laos. Consequently, General Lam had reassessed his plan of attack after the disappointing results of friendly armor in keeping open Highway Nine. Obviously he could no longer plan to use this as a secure main supply route. Capitalizing on his airmobile support, he decided to attack the main objective of Tchepone with a series of rapid air assaults along the high escarpment to the south of the river using the 1st Infantry Division.

From 3 to 6 March, the 1st Army of the Republic of Vietnam Division had accomplished a series of airmobile assaults to the west along the escarpment overlooking Route Nine. The first Army of the Republic of Vietnam units air assaulted successfully into landing zones LOLO, LIZ, and Fire Base SOPHIA WEST. After a very effective preparation of the area by B–52's, on 6 March two infantry battalions were lifted by 120 Hueys for 65 kilometers to air assault into landing zone HOPE north of Tchepone. *This large combat assault was carried out in what was considered to be the most hostile air defense environment ever encountered in the entire war, yet only one Huey was hit and it made a safe landing in the objective area.* The Army of the Republic of Vietnam units attacked south and west controlling the town. (*Map 12*) Tchepone was the objective of the allied drive to the west and was the natural communications hub of the enemy's logistics system in Laos. The enemy immediately increased his pressure in the Tchepone area and attacked the Army of the Republic of Vietnam fire bases on the escarpment viciously.

The I Corps Commander decided that most of the objectives of LAMSON 719 had been accomplished and ordered a timed withdrawal from Laos before weather worsened. During the extraction to the east from the Tchepone area, new enemy forces brought heavy pressure to bear on the Army of the Republic of Vietnam all along Route Nine. Extremely heavy antiaircraft fires were encountered along routes to or from the Army of the Republic of Vietnam fire bases. Enemy pressure was also felt at the primary U. S. Forward Support Area at Khe Sanh which received heavy

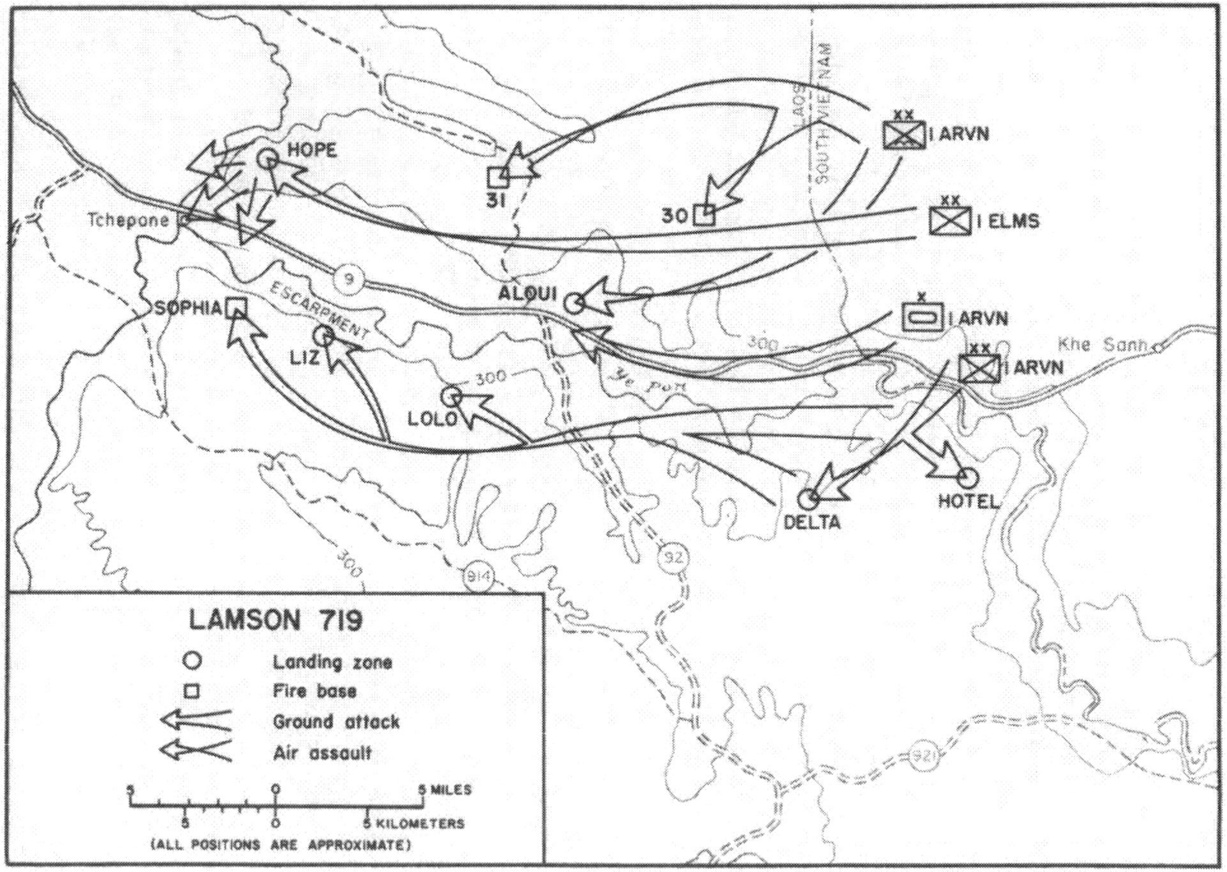
LAMSON 719
Landing zone
Fire base
Ground attack
Air assault
5 MILES
5 KILOMETERS
(ALL POSITIONS ARE APPROXIMATE)
HOPE
31
30
I ARVN
I ELMS
I ARVN
I ARVN
LAOS
SOUTH VIETNAM
Tchepone
9
ALOUI
SOPHIA
ESCARPMENT
LIZ
LOLO
300
Khe Sanh
DELTA
HOTEL
92
914
921

MAP 12

attacks by fire and sappers. All and all, the enemy used every means at his disposal to make the allied withdrawal as difficult as possible.

The last elements of the 1st Infantry Division were extracted on 21 March and the remaining Vietnamese forces withdrew back into South Vietnam over the next few days. The major airmobile actions in Laos were terminated by 25 March even though some Army of the Republic of Vietnam forces continued to operate across the border. Two highly successful airmobile raids of battalion size were conducted between 31 March and 6 April.

Thousands of tons of ammunition, petroleum, oils, and lubricants, and other supplies and equipment were destroyed by LAMSON 719 forces including U. S. air assets. In addition to the destruction of these stockpiles, supplies from the caches of Base area 604 were at least partially consumed by the North Vietnam Army forces opposing LAMSON 719. Initial reports of supplies and equipment destroyed or captured include over 4,000 individual weapons; more than 1,500 crew-served weapons; 20,000 tons of ammunition; 12,000 tons of rice; 106 tanks; 76 artillery pieces; and 405 trucks. The effectiveness of B–52 strikes, tactical air, helicopter gunships, and artillery is further indicated by over 9,700 secondary explosions.

As a minimum, it can accurately be stated that the enemy lines of communication in Base Area 604 were severed, and that supplies and equipment ceased to move south through this area during the inclusive dates of the operation. This was particularly significant, for in past years the enemy has reached his peak efficiency in moving resources south during the months of February and March. Additionally, the detailed knowledge obtained concerning the location of depots, trail networks, truck parks, and the fuel pipeline would permit more precise targeting in the future.

Enemy personnel losses were very heavy. While these losses might eventually be replaced, the requirement to replace losses in such regiments as the 1st Viet Cong, 29th, 36th, 64th, 102d, and 803d would, in all probability, draw off replacement personnel programmed for other units. Combined air-ground operations in Base Area 604 resulted in a reported total of 13,914 enemy killed in action. Air and ground attacks inside the five depot areas reportedly accounted for 5,357 of these casualties. An additional 69 enemy soldiers were captured.

Review of Airmobile Support During Lamson 719

The precise impact of LAMSON 719 on the enemy's long-range goals must be left for future studies. The important issue here is whether the airmobility concept failed or succeeded in this important test.

The average American citizen could not help but conclude from the headlines that the helicopter had proved to be an unacceptable combat vehicle. Many reporters picked up random stories from anyone willing to talk and the overall picture was grim. The following excerpt from *Newsweek*, 15 March 1971, was more objective than most:

> To the modern American cavalryman of the air, the plunge into Laos has been something like an old-time charge on horseback: admirably heroic, stunningly effective—and terribly costly. For four weeks now, American helicopter pilots have flown through some of the heaviest flak in the history of the Indochinese war. One day alone last week, the Army admitted to losing ten aircraft to the unexpectedly heavy North Vietnamese ground fire, and there were reports from the field that the actual losses had been much worse. As a result, the customary bravado of the American chopper pilot was beginning to wear a bit thin. "Two weeks ago," said one gunship skipper, "I couldn't have told you how much time I had left to serve in Vietnam. Now I know that I've got 66 days to go, and I'm counting every one." Another flier added anxiously: "The roles are reversed over there. In Vietnam, you have to hunt for the enemy. But in Laos, man, they hunt for you.
>
> Despite the risks, it was inevitable that U. S. helicopters should be deeply involved in the Laotian campaign, for more than any other artifact of war, the chopper has become the indelible symbol of the Indochina conflict. Helicopter pilots were among the first Americans killed in the war a decade ago, and, under President Nixon's Vietnamization program, they will probably be among the last to leave. In the years between, the chopper's mobility and firepower have added a radically new dimension to warfare, and the daring young American pilots have scooped up their Silver Stars, Distinguished Flying Crosses and Air Medals by the bushel—along with Purple Hearts. In the opinion of many military experts, the helicopter has been the difference between a humiliating U. S. defeat in Vietnam and whatever chance remains of attaining some more satisfactory outcome.

To put the story of airmobility during LAMSON 719 into focus, it's necessary to examine the threat. With the exception of enemy air, it could be said that the environment in Laos was as hostile and as sophisticated as most of the probable areas of employment of U. S. forces throughout the world. The North Vietnamese Army had skillfully deployed an extensive well-integrated, highly mobile air defense system throughout the entire operational area. Whereas in Vietnam and Cambodia we had operated against 7.62-mm and limited 12.7-mm fire, with occasional concentrations of the latter, operations in Laos had been regularly opposed by 23-mm, 37-mm and 57-mm weapons, while the 12.7-mm guns were employed in multiple mutual supporting positions. The enemy not only had large numbers of antiaircraft weapons of several calibers, but he used these weapons in a manner specifically designed to counter airmobile operations.

The North Vietnamese Army soldier enjoyed a considerably greater fire support in Laos than he had previously experienced in South Vietnam and his antiaircraft weapons had been carefully positioned over a period of years. The 12.7-mm weapons were often employed in triangular or rectangular formations in the vicinity of high ground approximately 1,000 meters from a potential landing zone. The 23-mm guns were employed in circular or triangular formations, though on occasion a single gun was used to protect storage sites or vital road networks. The extensive enemy threat was compounded by the fact that his antiaircraft weapons were continually redeployed, usually on a day-to-day basis.

One enemy tactic that proved most difficult to counter was the North Vietnamese Army technique of employing 10- to 12-man combat teams—on or near every piece of critical terrain—protected by bunkers and trenches. These small teams, armed with one or two machineguns and 82mm mortar and one or two rocket launchers, attacked allied aircraft and infantry on virtually every landing zone, pick up zone, and friendly troop position within the range of their weapons.

The enemy also used their "hugging" tactic which had proven effective in earlier encounters. Using this tactic, North Vietnamese Army forces sometimes moved to within 10 to 20 meters of friendly units manning perimeters and securing positions. Friendly forces were often reluctant to bring supporting fires close enough to their own positions to harm the enemy and, consequently, the close-in enemy could direct a heavy volume of short-range small arms, antiaircraft weapons, and rocket launcher fire against helicopters flying in and out of friendly positions. On occasion, helicopters were fired at and hit by North Vietnamese Army riflemen lying on and back *inside* of barb wire barriers surrounding a friendly position.

Because of the ever-present enemy threat, every airmobile operation in LAMSON 719—even single ship resupply and medical evacuation missions—had to be planned and conducted as a complete combat operation. This entailed a separate fire plan, allocation of escorting armed helicopters, and contingency plans for securing and recovering downed crews and aircraft.

The 101st Airborne Division (Airmobile), under the command of Major General Thomas M. Tarpley, was given the mission to provide support and assistance to the U. S. and Vietnamese forces participating in LAMSON 719 operations in western Quang Tri Province and in Laos, while still continuing the Division's winter campaign in Thua Thien Province. Furthermore, the Division would take over operational and security responsibility of the areas previously covered by the 1st Army of the Republic of Vietnam Division in Quang Tri Province and along the Demilitarized Zone. They would also conduct diversionary operations from the Hue area into the A Shau Valley along route 547. LAMSON 719 would receive top priority in all cases.

The support provided to I Corps Forces in Laos as well as to the U. S. forces operating in the northern provinces could not have been maintained at a high level throughout LAMSON 719 had not U.S. Army, Vietnam, devoted a major portion of its assets in support. Damaged or destroyed aircraft would be quickly replaced and maintenance support given priority to those aviation units assigned to or under the operational control of the 101st Airborne Division.

A special aviation task force organization was created to provide the extensive aviation support required by LAMSON 719. This aviation task force was built around the structure of the 101st Airborne Division (Airmobile) by supplementing the division's organic assets with aviation and air cavalry units from other divisions, the 1st Aviation Brigade, and from units scheduled for deactivation or redeployment. The division's 2d Squadron, 17th Cavalry, took operational control of supplemental air cavalry troops. The Division Support Command provided logistic and maintenance support for supplemental and organic units and established forward refueling and rearming points to support the operation. The 101st Division used its command and control structure to command the aviation and air cavalry units and to plan and conduct the airmobile operations in support of LAMSON 719.

One of the key U. S. commanders during LAMSON 719 was Brigadier General Sidney B. Berry, Jr. the Assistant Division Commander (Operations) of the 101st. He had a dual role as coordinator of U. S. aviation resources and defacto aviation officer to the Vietnamese I Corps Commander. These two hats made it possible for him to carry out the key position of senior commander aloft. In this position, he was separate from, and senior to, the ground and air mission commanders.

The availability of armed helicopters for the escort role was a major limiting factor in just how many different airmobile operations could be conducted simultaneously. To meet the demand for armed helicopters, many of the older UH-1C armed Hueys were committed to the action. However, this aircraft complicated the planning on armed helicopter support, for the older Huey could not keep up with, or perform as well as, the preferred Cobra.

With all its limitations, the armed helicopter proved the most important fire-support weapons system during LAMSON 719. Armed helicopters provided the capability for *detecting* and *immediately engaging* battlefield targets of opportunity close to friendly troops on the ground—a system unmatched by any other weapons system in the United States inventory. Armed helicopters, operating with the air cavalry, aerial rocket artillery, and escorting troop-lift, heavy-lift, and support aircraft, literally covered the battle area with their ability to respond immediately and accurately with their fire against known and suspected enemy weapons and positions. Armed helicopters often operated under low ceilings and weather conditions that restricted or precluded use of tactical air in close support of ground units or airmobile operations. Armed helicopters, particularly those of the air cavalry, played a key role in acquiring targets, directing artillery fire and tactical air strikes against them, and conducting battle damage assessments.

One can not overstate the importance of the air cavalry in this operation. They seemed to be omnipresent; they found most of the targets; and they were generally the first on the scene and the last to leave. All that I have said before about the merits of this organization was exemplified in Laos.

Air cavalry performed two principal missions during LAMSON 719—reconnaissance to the flanks and front of ground operations and reconnaissance and security of landing zones before and during combat assaults and extractions. Lieutenant Colonel Robert F. Molinelli, the Commanding Officer, 2d Squadron, 17th Cavalry was the principal reconnaissance officer for the operation.[1] Directing his four air cavalry troops, he took his assigned tasks directly from the Commanding General, I Corps, and delivered his reports back through both Army of the Republic of Vietnam and U.S. channels.[2] This system of assigning tasks and multiplicity of reporting channels testifies to the critical role played by the Air Cavalry. As the battle progressed it became evident that, because of their great confidence in the Air Cavalry, the Vietnamese units tended

[1] To be replaced by Lieutenant Colonel Archie A. Rider on 5 March.

[2] Colonel Molinelli put it this way: "I took my directions directly from General Lam, with General Sutherland supervising our operations on a daily basis. He (General Sutherland) was pretty much the on-the-scene commander until the combined tactical command post was established at Khe Sanh."

to employ the Air Cavalry in the close fire support role rather than in the reconnaissance role.

During LAMSON 719, the 2d Battalion 17 Cavalry encountered PT-76 tanks, a target new to the squadron.[3] Initially, anti-tank rockets were not available; engagement was made with ordnance on hand. Upon sighting a tank the Cobras would initiate contact at maximum range with 2.75-inch flechette rockets. This served to wipe personnel off the vehicles and their immediate proximity. As the gun run continued, the AH-1G pilots would begin firing a mixture of high explosive and white phosphorous rockets, breaking off the run at approximately 500 meters and, indeed, often overflying the target.

When available, the XM-35 20-mm cannon was used. This weapon was extremely accurate, and afforded a theoretical standoff distance of 2,000 to 2,500 meters; however, adequate ammunition was not available for this weapon. When high explosive, anti-tank rockets became available, results were mixed. This rocket was capable of penetrating armor, but direct hits on the target were required. Accuracy dictated that engagements be made at ranges of 500-1,000 meters from the target, thus exposing the gunship to the tank's 12.7-mm and to supporting enemy infantry in the area.

Upon sighting a tank or group of tanks, the Cavalry gunships would engage them to maintain contact, then normally turn the target over to the Air Force and continue reconnaissance missions. If Tactical air was not available, the gunships would engage tanks until their ordnance was expended; but they rarely had enough ordnance to destroy every tank in a particular sighting. Between 8 February 1971 and 24 March 1971, the Cavalry *sighted 66 tanks, destroyed* (*burned*) *six,* and *immobilized eight.*[4] Three of the destroyed tanks were hit with flechettes, High Explosive and White Phosphorous; and the other three were destroyed by combinations of flechettes, High Explosive, White Phosphorous, and High Explosive Antitank. The majority of the other tanks not destroyed or damaged by the Cavalry were turned over to the U. S. Air Force.

---

[3] The PT-76 cannot be truly classified as a "tank." It is better described as a lightly armored gun carriage. The Cavalry troops also sighted, but did not engage, T-34 tanks. In addition, there were reports (unconfirmed) from tactical air of an even heavier tank—the T-54.

[4] The following criteria were established by the 2d Battalion, 17 Cavalry to claim a tank destroyed or damaged. To classify a tank destroyed, the tank had to explode or burn; whereas a damaged tank was immobilized, parts were blown off and the tank was incapable of further movement without repair.

In summing up the exploits of his air cavalry troops, Colonel Molinelli said: "I think that the use of air cavalry in Laos pretty much proved two major points that those of us in Aviation have always maintained. One, our aircraft are not as vulnerable as many people think in a mid-intensity environment; two, we certainly proved that Army aircraft are capable of attacking enemy armor."

Describing the limitations of current armed helicopters, General Berry commented on 20 March 1971 as follows:

> . . . We need now tank-defeating armed helicopters. Had we entered LAMSON 719 with a helicopter armed with an accurate, lethal, relatively long-range anti-tank weapon, we would have destroyed many more NVA tanks and would have rendered more effective close support to Vietnamese ground forces. As I consider our experience against NVA tanks in LAMSON 719 and ponder what would face us on a European-type battlefield, I am absolutely convinced that the US Army must field immediately an armed helicopter with an effective tank-killing capability. If the AH–1G "Cobra" mounting the TOW gives us that required capability the soonest, fine. I hold no brief for any particular weapons system, but I do hold the firm conviction that we need now the armed helicopter tank-killer.

On this same date, General Berry commented on several other aspects of LAMSON 719. These comments—which were made before the actual close of the operation and right where the action was happening—bear repeating:

> . . . Our experience in conducting airmobile operations in support of LAMSON 719 confirms the soundness of the concept and principles of airmobility developed by the U.S. Army. We have, of course, modified and adapted specific tactics and techniques to cope with the operational environment. But airmobility principles and concepts have proven sound and valid.
>
> . . . Living and operating in the ground soldier's environment, the armed helicopter escorts troop-lift helicopters flying the soldier to and from his operations, escorts helicopters delivering ammunition, food, water, supplies, and mail to the soldier, and escorts the medical evacuation helicopter rescuing the wounded soldier from battle. The armed helicopter flies underneath ceilings measured in hundreds of feet to locate targets threatening or attacking the soldier to deliver timely, responsive, accurate fire within tens of feet of the soldier's position.
>
> The fighter-bomber has a unique capability to place heavy firepower and a variety of ordnance in close support of the ground soldier. The fighter-bomber's most distinctive characteristic is its ability to deliver heavy bombs in support of the ground soldier. The fighter-bomber flies underneath ceilings measured in thousands of feet, to deliver heavy bombs within hundreds of feet of the ground soldier's position and lighter ordnance even closer.
>
> The armed helicopter and fighter-bomber team works effectively in LAMSON 719. Armed helicopters of the air cavalry reconnoiter objective

areas, landing and pick up zones, and their approach and departure routes; acquire and mark targets on which the forward air controller directs air strikes; conduct low-level bomb damage assessments; and work with the forward air controller in developing additional targets for air strikes. Armed helicopters and tactical air work together to prepare the objective area, landing and pick up zones and approach and departure routes for safe passage and landing of the troop-lift helicopters. The armed helicopters then escort troop-lift and heavy-lift helicopters in and out of the landing zone while the forward air controller directs air strikes into adjacent target and danger areas.

. . . The helicopter and its crew have proven remarkably hardy and survivable in the mid-intensity conflict and hostile air defense environment of LAMSON 719. We have lost remarkably few helicopters and crew members in view of the heavy small arms, antiaircraft, and mortar and artillery fires our aircraft and crews have experienced while conducting extensive airmobile operations on NVA home ground. This is even more remarkable in view of the numerous airmobile operations conducted in support of Vietnamese ground units located in small perimeters, surrounded by NVA units and weapons, and often in heavy contact with the enemy.

To assess and evaluate properly our aircraft and crew losses, one must measure these losses against the campaign plan, mission, total sorties, and number of exposures to enemy fire, and accomplishments. When viewed in this perspective, we have fared better than the most optimistic prophet would have dared predict.

One of the great stories of LAMSON 719 is the magnificent effort to recover downed aircraft. Wherever possible, an immediate effort was made to extract any downed aircraft and crew, even in the face of hostile fire, and with the knowledge that the North Vietnames often used downed aircraft as bait with which to draw more equipment and personnel into an ambush. In some cases, the downed aircraft had to be destroyed because the tactical situation precluded recovery.

Research analysts will be working with the data base from LAMSON 719 for a long time, particularly on the vulnerability aspects of the helicopter. It would take several volumes to summarize their parameters and permutation alone. For the purpose of this monograph, I think it is fair to say that the loss rate experienced by Army helicopters compared favorably with the loss rate of high performance aircraft in Southeast Asia for the same period. Most importantly these loses were not considered unacceptable in view of the mission accomplished.

There is always the temptation to fall back on the trite statement that "war has never been a particularly safe business" and dismiss further discussion of loss in combat. However, no professional leader ever takes any of his losses lightly and the lessons learned from Laos will be studied intently for years to come to find better and safer means to carry out the airmobile mission. The general reaction of the Army aviator after LAMSON 719 was "if we could pull this off under these conditions, we can do it anywhere in the world." The senior commanders on the scene seem to share this conclusion.

In the context of the enemy's highly developed antiaircraft defense capability, can we make a valid judgment of the airmobile concept from the results of LAMSON 719? Let's be candid. Our total helicopter losses during this operation were 107 aircraft. Taken by itself, that figure seems a de facto indictment of the concept. But the last statistic does not tell the whole story—indeed, it is totally misleading if left unqualified.

The basic fact is: LAMSON 719 would never have been undertaken, much less successfully completed, without the support of thousands of helicopter sorties. *And for every thousand sorties the loss rate was only one quarter of one percent.* Granted, every helicopter loss was regretable; however, this ratio does show a very high rate of accomplishment versus attrition. Most of these losses were troop transport Hueys—and more than half of these were lost just as they approached landing zones. This again points out in the strongest way that the helicopter is most vulnerable as it comes to a hover over an unsecured or partially secured area. In other words, the safety of the helicopter pilot depends very much on the infantryman on the ground who he supports.

Not unexpectedly, the older Huey gunships did not fare as well as the Cobra in this intense air defense environment. The higher attrition rate of the armed Huey's proved that the move to the faster and better protected Cobra was timely and necessary. Many of the Cobras were hit by 12.7-mm fire but managed to return to base and, eventually, return to combat.

I recognize that this account of LAMSON 719 focuses on the airmobility aspects and does not pretend to tell the entire story of this important battle, a battle that is perhaps too recent to put into true historical perspective. One thing is certain. Without the air support of the U.S. Army, Air Force, and Marines, LAMSON 719 would never have even been planned, much less would it have succeeded.

CHAPTER XIII

Conclusions

The story of airmobility has been long—so my summary will be short. I've tried to interject my honest opinion where it seemed appropriate in the text. Only a few final observations are necessary.

For the reader who has borne with me through this account of ten years of airmobility in Vietnam and the fundamental decisions in the decade preceding that, the obvious question is, "What does all of this mean?" The one inescapable conclusion is that the airmobility concept is irreversible. The thousands of officers who have learned to think and fight and live in three dimensions will never allow themselves to be restricted to two dimensions in the future. Airmobility will change and grow, but it is here to stay.

In the first chapter we learned that the growth of the airmobile concept did not take place in the framework of guerrilla warfare. It was conceived out of the necessity to disperse on the modern battlefield under the threat of nuclear weapons and still retain the ability to mass quickly for decisive actions, then disperse again. The actualities of Vietnam have since obscured these origins and have led many people to the assumption that airmobility was designed for and limited to counter-guerrilla contingencies. The very nature of the terrain in Vietnam with its jungles and mountains has led many to connect helicopter operations to this type of terrain. Indeed, the opposite is true. Airmobility worked in Vietnam *in spite* of the tremendous problems of working in the jungles and the mountains of an undeveloped country. The helicopter overcame the obstacles of limited landing zones, primitive road nets, restricted observation, and high density altitudes as no other vehicle could. But, in the open countryside of Europe or a desert in the Middle East, the airmobile force has far greater flexibility and many more options than even the armored forces of Rommel in North Africa. Vietnam represented only a fraction of the possibilities for airmobile tactics.

A casual observer of Army aviation in Vietnam could easily have arrived at the conclusion that there was no need for specialized airmobile divisions. In the latter years, especially 1967 and 1968, there were enough aviation assets to satisfy almost every requirement for airmobility in every division. Many of the non-airmobile divisions, and rightly so, were very proud of their air assault techniques and had developed highly refined operating procedures with their supporting aviation companies. The question then arises as to the need for a special organization which has over 400 helicopters assigned to it on a full-time basis.

However, there is much more to airmobility than just using helicopters. The Huey, the Cobra, the light observation helicopter, and the Chinook were the essential vehicles of airmobile combat and combat support. But, the essence of an airmobile division is the ability to integrate the capabilities of these aircraft into an organization designed for their use by people specially trained for their use. This total integration is only possible in a unit which "owns" its helicopters.

This is not to detract from the effectiveness of non-airmobile units for they possess certain characteristics which give them advantages in certain situations. A division which has tank or mechanized battalions or ground cavalry squadrons has certain advantages over the airmobile division. The same thing is true for the heavier weapons which are organic to the infantry division. Conversely, the lighter weapons and vehicles of the airmobile division are essential for the advantages which it possesses. This study has detailed the strategic moves of the 1st Cavalry Division from II Corps to I Corps to III Corps Tactical Zones, and touched upon the many brigade and battalion moves involving airlift. In all these moves, I was struck by the distinct advantage in moving inherently *lean* airmobile forces as compared to the airlift requirement of other organizations.

The tactical advantages of the airmobile division can be summed up as follows: increased efficiency due to the repeated association of units; thorough integration of its assets because of close association and command relationships; and the ability to take a different conceptual approach because of its assured assets. The impact of organizational and command relationships has a direct and distinct impact on the quality of support. There is no denying that general support units rarely tend to identify closely with the supported unit, at least not as closely as organic units. This is a simple truism of human nature.

The epitome of "superior" airmobile tactics, as contrasted with "good," is the capability to exploit not only an opportunity but the trends and changing patterns in enemy activity. This requires a flexibility which is much more difficult to initiate and accomplish when non-divisional units are involved. This is particularly true of air cavalry. More than any other unit, the air cavalry development in Vietnam has proven the need for quick reaction to meet the demands of the situation as seen by the commander on the scene. Adaptation to a change in environment is more feasible and more likely when the unit is organic.

The airmobile division commander is able to plan a complete campaign based on airmobility as opposed to a campaign which, as an incidental element, employs airmobile assets in some of its battle plans. As a corollary, I know of no other major organization besides the airmobile division where the commander is willing to consistently commit all his forces on a day-to-day basis; that is, not keep the traditional reserve. The airmobile division commander *knows* that he can extract a reserve out of *his* deployed units as needed because he has the assets and the training to do so. This is a major economy of force.

By changing our perspective from that of the division commander to the individual soldier, we can perceive one of the often forgotten advantages of airmobility. It is ironic to me, after the millions spent to reduce a few pounds from the infantryman's rifle, that many soldiers would end up carrying as much (or more) total weight as the doughboy of World War I. A soldier when he exits a helicopter becomes the Army's most important extension of the airmobile concept; and his individual load should consist of the bare essentials needed for the next few hours—basically ammunition and water. To avoid the classic soldier's syndrome of holding on to everything he owns, the 1st Cavalry Division developed a technique for keeping all the personal equipment of the individual in squad bundles that were consistently delivered when needed. This was a tremendous boost to morale as well as a very real increase in effectiveness. However, this seemingly simple technique has to be relearned several times in every campaign.

In reviewing this volume, I sense that I could have spent more time emphasizing the natural affinity of armor with an airmobile force. When the terrain and circumstances permit, armor and airmobility complement each other in a natural way to form an unbeatable team. Airmobility gives the commander unique capabilities in reconnaissance, maneuver, and logistics while the armor gives the shock and firepower which have characterized it in the past. Air cavalry and airmobile infantry can find and fix the enemy so that armored and mechanized forces can be brought in at the decisive moment to finish him. There is no precise balance of these forces that would apply across the board. It will be up to the theater commander and subordinate commanders to mix these two elements to form the specific compound for a particular situation.

Many new organizations will be conceived as a result of our Vietnam experience and that is as it should be. However, we must make certain we do not invent something we have already thoroughly tested in combat. For example, there is a great deal of experience in the record of armor working with airmobile units; there was an air cavalry combat brigade, in fact if not in name, operating as part of the 1st Cavalry Division in 1970. We should be certain that we pick up any extrapolations for the future where we left off in combat. There is so much work that needs to be done that we can ill afford time to prove what has been proven or preparing answers to questions that will not be asked again.

We must use a similar approach to future Army–Air Force relationships. One lesson should stand out loud and clear from our ten years of experience in Vietnam: the command and control procedures evolved in combat, often hammered out by the very men whose lives depended on them, proved sound and workable. While this monograph has not tried to detail the Air Force story in Vietnam, I trust it has given enough examples of the magnificent support that the Air Force provided, and the trust and confidence that was generated in the minds of every major ground commander. As our experience grew, the close integration and timing of Air Force support to the organic Army support could not have been improved. I see no need for the Army and the Air Force to go through another agonizing reappraisal of their command and control structures (and the haggling over hardware to do the job) such as was experienced after World War II and Korea. In these latter two periods, valuable and important lessons were forgotten in the peacetime budget exercises while the Services engaged in bitter and often emotional debates which proved unnecessary and detrimental. Now is the time to capitalize on the vast reservoir of experience in both Services to put the ad hoc arrangements of Vietnam into lasting doctrine. It is not the time to debate new interfaces, new organizations and new command relationships that are untested.

In any activity where two Services operate the interface between them will seldom be a comfortable enmeshing of capabilities—whether these capabilities are competitive, supplementary, or complementary. Army–Air Force aviation relationships are no exception. The important thing to recognize is that there are gaps more often than overlaps in all areas of the interface. This has occurred because each Service is able to provide less of a capability, at least in qualitative terms, than is militarily desirable. This is especially true as we look to more sophisticated threats throughout the world. In my estimation, quantitative duplication is a myth, because seldom will relatively scarce and expensive vehicles such as aircraft be adequate to meet all requirements.

Another myth—one that has haunted Army aviation for thirty years—is vulnerability. Throughout this volume we have touched on various aspects of the vulnerability of Army aircraft and, in particular, the helicopter. Nevertheless, the subject is still uppermost in the minds of those who have honest reservations on the soundness of future airmobile tactics.

When I began the research for this study, I sent letters to fifty senior officers requesting their personal comments and experiences on the subject of airmobility. Many of their thoughts have been incorporated in this volume. Only one letter, from Lieutenant General Arthur S. Collins, Jr., voiced serious doubts about the future of the airmobile concept. In his detailed analysis, which is an important part of the bibliography, General Collins listed nine major strengths of the airmobile concept; however, he still believed that "the fundamental weakness of the helicopter, and therefore the entire airmobile concept, is its vulnerability to ground-to-air fire." I disagree. But such concern cannot be lightly dismissed. Those charged with the responsibility of planning our future force structure must weigh very carefully their decisions on the type and amount of airmobile organizations against the possible contingencies. In the preface to this volume I stated that airmobility was no panacea. Some of the operations in Vietnam showed that there were higher risks involved for higher gains. I submit that the entire strategy in Southeast Asia would have had to be revised downward if we had not had airmobile organizations, imperfect though they were.

The key word for airmobile operations is "survivability," not "vulnerability." Survivability of air vehicles in the land battle is one end product of a combination of actions and reactions by two opposing forces. The kinds of battlefield actions and reactions are many and varied, beginning with intelligence production and planning and ending with the last shot fired. Survivability of aircraft can be appreciated only by examining all of these influences in their proper relationship to each other.

The oft-studied subject of vulnerability, which is *only one* input to survivability, has to be recast in proper perspective with regard to other equally or more important contributors. The development of this perspective requires considering similarities which exist among the survivabilities of all combat elements to include Army air vehicles. The survivability of Army aircraft is enhanced by suppressive ground fire support, close air defense support, the proper use of intelligence for planning aviation operations, the effect of tactic and techniques on increased survivability, the soldier's desire to accomplish his mission, and the effect of personal command attention. Since Army aircraft operate in the ground environment, proven techniques of ground survival are available to them; and, the most effective of these techniques is the co-ordinated use of all his capabilities by a commander on the scene. What is germane is the fact that the American soldier is more capable of carrying out his mission and more likely to survive in combat because he is airmobile.

Throughout this study we have tried to emphasize the "man" portion of the man–machine equation. However, I'd like to reiterate a few of the equipment lessons that seem exceedingly important during future procurement decisions. History has proven that the tools of combat are seldom if ever, used in the exact manner which their designer had conceived. For example, the Huey began life as an air ambulance. Therefore, it is essential that today's developer, in laying the keels for the next generation, design the most *useful* tools, so that some future commander will have available a flexible arsenal that he can adapt to his war.

The Army's decision to standardize on a utility tactical transport helicopter has far-reaching implications on every operation from its planning to its execution. Literally hundreds of our key battles could not have been fought without a light, agile machine that could go into improbable landing zones at a critical time. Had the Army chosen to build its airmobile tactics around a "platoon carrier," different and less flexible tactics would have been forced on our commanders. As we move to replace the Huey fleet, we must never lose sight of the essential characteristics that made the Huey invaluable to the Infantry commander. Technology offers so many tempting alternatives that one can easily forget the basic problems of squad tactics. The vital lessons which we learned in the "sizing" of our helicopter fleet dare not be forgotten.

Similar principles must be kept in mind in our future helicopter gunships. There is no doubt that from a hardware standpoint, the armed helicopter was the single greatest innovation of Vietnam. I trust this volume has provided enough examples of its unique characteristics and methods of employment to leave no doubt about its place in the support of the ground soldier. The introduction of the Cobra into Vietnam vindicated all the hypotheses of the armed helicopter pioneers who were derided in their early experiments. The Cobra came at the right time in sufficient numbers to do a job that no other fire support means could do. LAMSON 719, if it proved any point at all, proved that the Cobra could survive under high intensity warfare while the older Huey gunship merely showed deficiencies that we knew had always existed.

But, there is another story in this volume that has perhaps not been emphasized enough, and that is the test of the "Go-Go Bird" or armed Chinook. Now, I am speaking of a frame of mind that wanted to produce a "battleship" (with all the firepower that term implies) rather than an agile "destroyer" with the agility to go in and get out. For the future, I think it is possible to reap the benefits of the latest technology in weapons systems without producing another "battleship" with its inherent disadvantages.

The story of airmobility in Vietnam is almost certainly just the first chapter of a new and dynamic Army. The glamour of airmobility has long passed, but the challenges are as great as ever. Some of the technological forecast, just dimly seen by the early planners, is now reality. If this study has served any purpose besides its bibliography, which I think is most important, it will form part of the corporate memory for those planners of the future who would like not to pay the terrible price of relearning again in combat many costly lessons. As the poet-diplomat Paul Claudell once observed, "It is not enought to know the past, it is necessary to understand it."

APPENDIX
ARMY AIRCRAFT PHOTOGRAPHS

UH-19D Chickasaw

Mission: utility, including transportation of cargo and troops, rescue, and observation operations

First delivery: 1953

Rotor diameter: 53′

Length: 62′3″

Basic weight: 5,650 lbs.

Payload: 844 lbs.

External cargo: 2,000 lbs.

Crew: 2

Passengers: 10

Cruise airspeed: 70k

Maximum airspeed: 115k

None remaining in active Army

CH–34C Choctaw

Mission: transportation of cargo and personnel

First delivery: 1955

Rotor diameter: 56′

Length: 65′10″

Basic weight: 7,800 lbs.

Payload: 2,175 lbs.

External cargo: 3,000 lbs.

Crew: 2

Passengers: 18

Cruise airspeed: 85k

Maximum airspeed: 133k

230 currently in Army inventory

CH–23 Raven (D&G models)

Mission: reconnaissance, medical evacuation, pilot training

First delivery: 1955

Rotor diameter: 35′5″

Length: 40′8 ½″

Basic weight: 1,821 lbs.

Payload: 851 lbs. (D model) ; 600 lbs. (G. model)

Crew: 1

Passengers: 2

Cruise airspeed: 70k

Maximum airspeed: 83k

941 currently in Army inventory

OH–13S Sioux

Mission: visual observation and target acquisition, pilot training

First delivery: 1956

Rotor diameter: 37′ 1 ½″

Length: 43′3″

Basic weight: 1,715 lbs.

Payload: 400 lbs.

Crew: 1

Passengers: 2

Cruise airspeed: 70k

Maximum airspeed: 87k

575 currently in Army inventory

CH–37B Mohave

Mission: transportation of cargo, equipment, and troops

First delivery: 1956

Rotor diameter: 72′

Length: 88′

Basic weight: 21,500 lbs.

Payload: 5,300 lbs.

External cargo: 10,000 lbs.

Crew: 3

Passengers: 23

Cruise airspeed: 90k

Maximum airspeed: 110k

17 currently in Army inventory

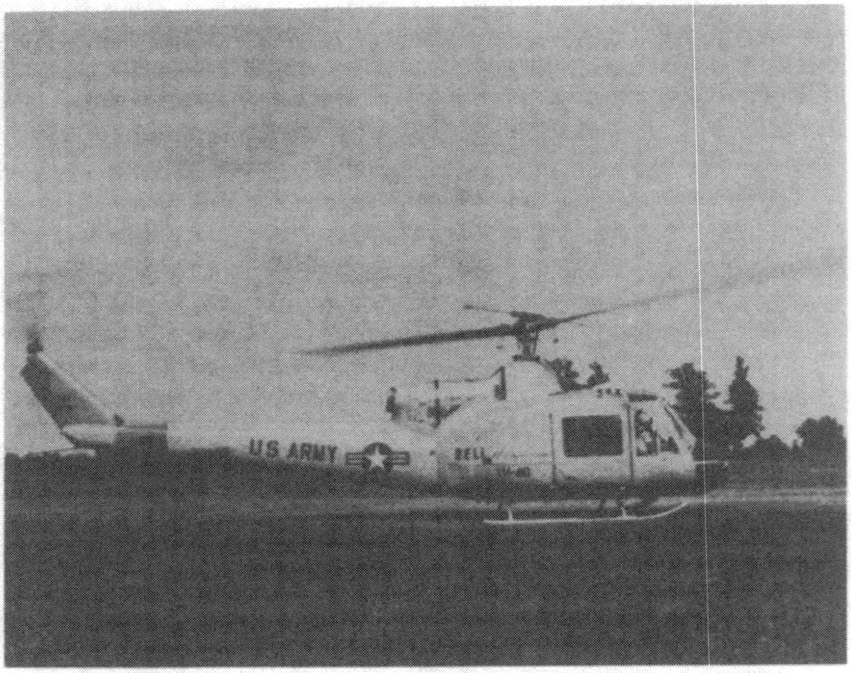

XH–40

Prototype for UH–1A

Same specifications as UH–1A

None in Army inventory

UH–1A Iroquois

Nickname: Huey

Mission: transport of personnel, equipment, supplies, training aircraft

First delivery: 1959

Rotor diameter: 43′9″

Length: 52′10″

Basic weight: 4,020 lbs.

Payload: 2,175 lbs.

External cargo: 3,000 lbs.

Crew: 2 Passengers: 6

Cruise airspeed: 80k

Maximum airspeed: 120k

54 currently in Army inventory

CH–47 CHINOOK (A,B,&C MODELS)

Mission: transportation of cargo, equipment and troops

First delivery: 1961

Rotor diameter: 59′1 ¼″ (A&B) ; 60′ (C)

Length: 98′ 3 ¼″ (A&B) ; 99′ (C)

Basic weight: 18,500 lbs. (A) ; 19,194 lbs. (B) ; 19,772 lbs. (C)

Payload: 10,114 lbs. (A) ; 15,900 lbs. (B) ; 19,100 lbs. (C)

External cargo: 16,000 lbs. (A) ; 20,000 lbs. (B) ; 22,700 lbs. (C)

Crew: 3 Passengers: 33

Cruise airspeed: 110k (A) ; 140k (B) ; 150k (C)

Maximum airspeed: 130k (A) ; 165k (B&C)

Armament: 7.62-mm door-mounted machine gun

482 currently in Army inventory

CH–21C SHAWNEE

Mission: transportation of cargo, equipment, and personnel

First delivery: 1962

Rotor diameter: 44′

Length: 86′4″

Basic weight: 8,900 lbs.

Payload: 1,920 lbs.

External cargo: 5,000 lbs.

Crew: 2

Passengers: 20

Cruise airspeed: 80k

Maximum airspeed: 120k

None remaining in active Army

UH-1B Iroquois

Nickname: Huey

Mission: transport of personnel, equipment, supplies, and to serve as an aerial weapons platform

First delivery: 1961.

Rotor diameter: 44′

Length: 52′11″

Basic weight: 4,600 lbs.

Payload: 2,704 lbs.

External cargo: 4,000 lbs.

Crew: 2–4 Passengers: 8

Cruise airspeed: 90k Maximum airspeed: 120k

Armament: 40-mm grenade launcher, 7.62-mm machine gun, 2.75″ rockets, M22 guided missile

456 currently in Army inventory

UH–1D Iroquois

Nickname: Huey

Mission: transportation of personnel, equipment, and supplies, medical evacuation, delivery of protective fire by attachment of appropriate weapons, and instrument training

First delivery: 1963.

Rotor diameter: 48′3″

Length: 57′1″

Basic weight: 4,900 lbs.

Payload: 3,116 lbs.

External cargo: 4,000 lbs.

Crew: 2–4

Passengers: 11

Cruise airspeed: 100k

Maximum airspeed: 120k

Armament: 7.62-mm door-mounted machine guns

1,010 currently in Army inventory

UH–1H Iroquois

Same as UH–1D, but with more powerful engine

2,399 currently in Army inventory

UH-1C Iroquois

Nickname: Huey

Mission: transporting personnel, special teams or crews, equipment, and supplies; medical evacuation; ambulance service; reconnaissance and security; point target and area fire by attachment of appropriate weapons; and instrument trainer

First delivery: 1965

Rotor diameter: 44′

Length: 53′

Basic weight: 4,830 lbs.

Payload: 4,500 lbs. External cargo: 4,000 lbs.

Crew: 2–4 Passengers: 6

Cruise airspeed: 100k Maximum airspeed: 140k

Armament: 40-mm grenade launcher, 2.75″ rockets, minigun

290 currently in Army inventory

CH-54 TARHE

Nickname. Sky Crane

Mission: movement of heavy outsized loads, recovery of downed aircraft, and by use of detachable pods, transportation of personnel, vehicles, and equipment

First delivery: 1966

Rotor diameter: 72′

Length: 88′5″

Basic weight: 20,700 lbs.

Payload: 15,400 lbs.

External load: 20,760 lbs.

Crew: 3 Passengers: 0

Cruise airspeed: 100k Maximum airspeed: 130k

74 currently in Army inventory

OH–6A Cayuse

Mission: visual observation and target acquisition, reconnaissance, command and control

First delivery: 1966

Rotor diameter: 26′4″

Length: 30′3″

Basic weight: 1,157 lbs.

Payload: 930 lbs.

Crew: 1

Passengers: 3

Cruise airspeed: 100k

Maximum airspeed: 130k

Armament: minigun

613 currently in Army inventory

AH–1G Huey Cobra

Mission: enroute escort reconnaissance, direct fire support

First delivery: 1967

Rotor diameter: 44′

Length: 52′ 11 ½″

Basic weight: 5,783 lbs.

Payload: 1,993 lbs.

Crew: 2

Passengers: 0

Cruise airspeed: 130k

Maximum airspeed: 190k

Armament: 40-mm grenade launcher, minigun, 2.75″ rockets, guided missiles

568 currently in Army inventory

OH–58A Kiowa

Mission: visual observation and target acquisition, reconnaissance, command and control

First delivery: 1969

Rotor diameter: 35′4″

Length: 32′3 ½″

Basic weight: 1,583 lbs.

Payload: 760 lbs.

Crew: 1

Passengers: 4

Cruise airspeed: 102k

Maximum airspeed: 120k

Armament: minigun

751 currently in Army inventory

O–1 Bird Dog (Formerly L–19)

Mission: reconnaissance and observation

First delivery: 1950

Wing span: 36′

Length: 25′

Basic weight: 1,542 lbs.

Payload: 100 lbs.

Crew: 1

Passengers: 1

Cruise airspeed: 86k

Maximum airspeed: 101k

552 currently in Army inventory

U–1 Otter

Mission: utility transport, light cargo hauling, passenger service

First delivery: 1950

Wing span: 58′

Length: 41′10″

Basic weight: 4,900 lbs.

Payload: 1,398 lbs.

Crew: 2

Passengers: 10

Cruise airspeed: 104k

Maximum airspeed: 137k

112 currently in Army inventory

U–6 Beaver

Mission: utility transport, light cargo hauling, passenger service

First delivery: 1951

Wing span: 48′

Length: 30′5″

Basic weight: 3,100 lbs.

Payload: 930 lbs.

Crew: 1

Passengers: 5

Cruise airspeed: 105k

Maximum airspeed: 142k

503 currently in Army inventory

U–8 Seminole (D&F models)

Mission: transportation of personnel

First delivery: 1956

Wing span: 45′3″ (D); 45′10 ½″ (F)

Length: 31′6″ (D); 33′4″ (F)

Basic weight: 4,981 lbs. (D); 5,381 lbs. (F)

Payload: 425 lbs. (D); 830 lbs. (F)

Crew: 1

Passengers: 4

Cruise airspeed: 155k (D); 160k (F)

Maximum airspeed: 200k (D&F)

172 currently in Army inventory

C–7 Caribou (formerly CV2)

Mission: cargo hauling, passenger transport

First delivery: 1959

Wing span: 95′8″

Length: 74′

Basic weight: 20,000 lbs.

Payload: 5,000 lbs.

Crew: 3–4

Passengers: 32

Cruise airspeed: 165k

Maximum airspeed: 210k

4 currently in Army inventory

OV–1 MOHAWK (A,B,&C MODELS)

Mission: visual reconnaissance and observation, photographic reconnaissance (A model), electronic surveillance using side-looking airborne radar (SLAR) (B model), Infrared (IR) reconnaissance (C model)

First delivery: 1961

Wing span: 42′ (A,C) ; 48′ (B)

Length: 41′1″

Basic weight: 9,781 lbs.

Payload: 341 lbs. (A&B models) ; 741 lbs. (C model)

Crew: 2

Cruise airspeed: 185k

Maximum airspeed: 255k

275 currently in Army inventory

U–21 UTE

Mission: utility services, command control, administration, liaison, aeromedical evacuation

First delivery: 1967

Wing span: 45′10 ½″

Length: 35′0″

Basic weight: 5,250 lbs.

Payload: 1,600 lbs.

Crew: 2

Passengers: 10

Cruise airspeed: 210k

Maximum airspeed: 230k

95 currently in Army inventory

Glossary

ACR	Air Cavalry Regiment
AIR ASSAULT II	The major and final Army test of the 11th Air Assault division in the Carolinas.
AKA	Attack Cargo Ship
AO	Area of Operations
ARA	Aerial Rocket Artillery
ARVN	Army of the Republic of Vietnam
ASOC	Air Support Operations Center
ATTLEBORO	A battle focused around the Special Forces camp at Sui Da, 19 October–24 November 1966.
Bde	Brigade
BLACKHAWK	A task force made up of the 7th Battalion, 1st Cavalry in support of the 44th Special Tactical Zone.
Bn	Battalion
BYRD	A battalion size microcosm of the 1st Cavalry Division's operations in Binh Dinh province.
Cav	Cavalry
CC	Command and Control
CHOPPER	The first airmobile combat action in Vietnam, commencing in December 1961.
Co	Company
COUGAR	A task force made up of the 214th Aviation Battalion in support of the 9th and 7th Army of the Republic of Vietnam Divisions.
CRAZY HORSE	An operation in defense of the Vinh Thanh Civilian Irregular Defense Group Camp in Binh Dinh province.
CRID	Capitol Republic of Korea Infantry Division

CRIMP	A 173d Airborne drive through the Ho Bo Woods region of Binh Duong Province in an attempt to destroy the politico-military headquarters of the Viet Cong Military Region 4.
CTZ	Corps Tactical Zone
DAN CHI 157	An engagement with the Tay Do Battalion in Can Tho Province in the Delta during which the 13th Aviation Battalion won the Presidential Unit Citation.
DASC	Direct Air Support Center
DAVY CROCKETT	A 3d Brigade operation covering the northeast portion of Binh Dinh Province, 4–16 May 1966.
DELAWARE–LAM SON 216	The 1968 operation in the A Shau Valley.
DMZ	Demilitarized Zone
EAGLE THRUST	Deployment of the balance of the 101st Air Cavalry Division, 8–18 December 1967; the largest and longest military airlift ever attempted into a combat zone.
FAC	Forward Air Controllers
G–1	Division Assistant Chief of Staff for Personnel
G–2	Division Assistant Chief of Staff for Military Intelligence
G–3	Division Assistant Chief of Staff for Operations
G–4	Division Assistant Chief of Staff for Logistics
G–5	Division Assistant Chief of Staff for Civil Affairs
GOLD FIRE I	A Strike Command sponsored joint test and evaluation exercise conducted in Missouri in 1964.
GUARDIAN	A task force made up of the 13th Aviation Battalion in support of the 21st Army of the Republic of Vietnam Division.

HAWTHORNE	An operation in Kontum Province.
HIGHLAND	Movement of the 1st Brigade, 101st Airborne Division to secure An Khe, 22 August–2 October 1965.
Hq	Headquarters
IRVING	A 1st Cavalry Division operation in Binh Dinh province in 1966.
JEB STUART	1st Cavalry operation in I Corps, including the move north, the *Tet* offensive, securing base areas, and preparation for PEGASUS.
JP-4	Kerosene-based fuel for turbine engines
JUNCTION CITY	A major operation north of Tay Ninh City in 1966.
JUNCTION CITY ALTERNATE	An operation begun on 22 February 1967 with the only parachute assault in Vietnam.
KBA	Killed By Action
KIA	Killed In Action
LAM SON 719	A combined operation in Laos from 8 February to 6 April 1971.
LCM	Landing Craft, Mechanized
LEJEUNE	A 1st Cavalry Division operation to relieve the Marines in Quang Ngai Province, 7–22 April 1967.
LIBERTY CANYON	The move of the 1st Cavalry Division from I Corps to III Corps in the fall of 1968.
LPD	Amphibious Transport Dock
LSD	Landing Ship, Dock
LST	Landing Ship, Tank
LZ	Landing Zone
MACV	Military Assistance Command, Vietnam
MARAUDER	An operation conducted in the Plain of Reeds of the Delta by the 173d Airborne Brigade, 1–8 January 1966.
MARKET TIME	U.S. Navy operations offshore of Vietnam under Task Force 115.

MASHER	The first phase of the initial major operation of the 1st Cavalry in Binh Dinh Province, 1966.
MATADOR	An operation to find and destroy the enemy in Pleiku and Kontum Provinces, conducted during the first half of January 1966 by the 1st Brigade of the 1st Cavalry Division.
NATHAN HALE	An operation in Phu Yen Province.
NET	New Equipment Training
NEVADA EAGLE	A 101st Air Cavalry Division operation, a large rice-denial effort, in the plains south of Hue.
NPFF	National Police Field Force
NVA	North Vietnam Army
OPCON	Operational Control
OSD	Office of the Secretary of Defense
PAUL REVERE	An operation in Pleiku Province.
PAUL REVERE IV	An operation near the Cambodian border in Pleiku Province.
PEGASUS	The operation to relieve Khe Sanh, 1–15 April 1968.
PERSHING	A 1st Cavalry Division year-long operation in Binh Dinh Province in 1967, terminating 21 January 1968.
ROAD	Reorganization Objectives Army Divisions
RVN	Republic of Vietnam
RVNAF	Republic of Vietnam Armed Forces
S–3	Officer in charge of the operations and training section of a brigade or smaller unit.
SAR	Search and Rescue
SAW	Special Air Warfare
SOP	Standing Operating Procedure
TAC	Tactical Air Command
TAOR	Tactical Area of Responsibility

THAYER I	A 1st Cavalry Division operation in Binh Dinh Province, 15–30 September 1966.
THAYER II	A 1st Cavalry Division operation in Binh Dinh Province in 1966.
TOAN THANG 43, 45, and 46	Operation in the "Fishhook" of Cambodia
TOC	Tactical Operations Center
TOW	An anti-tank wire-guided missile
TRUONG CONG DINH	A 9th Infantry Division operation in Dinh Tuong and Kien Tuong Provinces in the IV Corps.
VC	Viet Cong
VINH LOC	A combined operation of elements of the 101st Airborne Division, the U.S. Navy, and Republic of Vietnam forces conducted in Thua Thien Province, 10–20 September 1968.
VTOL	Vertical Take-off and Landing
WALLOWA	A 1st Cavalry Division operation between Chu Lai and Da Nang in I Corps beginning 4 October 1967.
WHITE WING	The second, third, and fourth phases of the initial major operation of the 1st Cavalry in Binh Dinh Province, concluded 17 February 1966.

Index

222-408

PIN : 039658-000

Made in the USA
Monee, IL
07 November 2026

68002974R00177